T0259044

INNOVATION
CRISIS

INNOVATION CRISIS

CRISIS

Successes, Pitfalls, and Solutions in Japan

Eiichi Yamaguchi

PAN STANFORD PUBLISHING

Published by

Pan Stanford Publishing Pte. Ltd.
Penthouse Level, Suntec Tower 3
8 Temasek Boulevard
Singapore 038988

Email: editorial@panstanford.com
Web: www.panstanford.com

British Library Cataloguing-in-Publication Data
A catalogue record for this book is available from the British Library.

ISBN 978-981-4774-97-0 (Hardcover)
ISBN 978-0-429-44862-1 (eBook)

Contents

Preface

This book is a summary of my research over the last 20 years of my life, especially after I moved out of the field of physics research to which I had devoted 21 years. I put in my own savings and bought the huge experiment device, which I designed and built when I was in the south of France, to enable me to dedicate the rest of my life to physics research. I felt that if I did not do my bit to save this sinking ship, it would be disrespectful to the future generation.

What has gone wrong in Japan where both innovation and science are simultaneously perishing? Let us go back to the source of innovation and find out what is wrong. Although my awareness about this issue started off from this point, it spanned several academic disciplines, and it appeared as though research had no context.

However, everything was connected, a fact that was confirmed to me when I finished writing.

Only after putting all the different pieces together did I realize that it was actually one story. In fact, the theory of innovation centered around the innovation diagram, and the theory of trans-science found by generalizing the TEPCO (Tokyo Electric Power Company) Fukushima Daiichi nuclear accident actually have the same root.

I would like to express my sincere thanks to journalist Yoshihiro Kataoka. He looked at my entire work, and put things into perspective.

A part of this book was backed by the research program "Science of Science, Technology and Innovation Policy" of the Japan Science and Technology Agency (JST). I really appreciate their support.

I am also thankful to my son-in-law, Alexander Kaymak for English editing. He did his best to improve my English, but it would be my fault if there should remain any errors in English.

Last but not the least, I am extremely thankful to my daughter Eri Kaymak who tended to me every single day despite being in her last month of pregnancy. I would like to dedicate this book to their children, George and Lily Kaymak, who came into this world while I struggled for writing.

Eiichi Yamaguchi
Kyoto, Japan
January, 2019

Prologue
How blue LEDs led to Nobel Prize

1. Case of Akasaki and Amano

At Nagoya University, Isamu Akasaki (1929–present, Photo 1) was impressed by the transparent wafer that a graduate school student, Hiroshi Amano (1960–present, Photo 2) showed him.

Photo 1 Isamu Akasaki (taken at Meijo University on December 29, 2000)

Photo 2 Hiroshi Amano (taken at Meijo University on March 13, 2001)

Four years had passed since Akasaki had left the Tokyo Research Institute at Matsushita Electric (current Panasonic) and had gone back to working for Nagoya University in 1981. Under the advice of Akasaki, Amano had persisted in his research on attempting to grow gallium nitride (GaN) single crystals on the surface of sapphire wafers on a system called MOVPE (metalorganic vapor phase epitaxy). Over 1500 experiments had ended in failure, leaving him with opaque, strange looking creations each time. When he inspected these creations with a microscope, he always found that the surface looked like uneven, six-sided polygons overlapping one another. To him, it looked more or less like the surface of the moon.

However, one day, Akasaki finally saw something that looked completely different from all the others. In one instance, it seemed as though the sapphire hadn't exhibited the growth of GaN at all, and it appeared to be clear.

Akasaki placed the wafer in a position facing the light and, from a sidelong view, saw the reflecting light from the surface. When the light was filtered and split into a rainbow, there appeared to be some kind of film visible from the surface, as well as some other hue mixing in with the other seven colors.

He was able to tell that the GaN was indeed growing. There was a flawless single crystal on the mirror surface. After observing this wafer for a while, Akasaki stared at Amano, who was hovering around, and said, "Try to measure the degree of crystallization using X-ray diffraction." Amano could see that Akasaki had finally realized his dream. That momentous breakthrough came through on one otherwise mundane evening in February of 1985.

The next morning, Amano got on an Osaka-bound train at Nagoya Station, to visit Osaka Prefecture University, which had an X-ray diffusion meter. While on the train, he gazed discreetly at the GaN crystals he had finally succeeded in making. Previously, whenever he had brought his creations to show his professor, he would always be told, "It's still not quite there yet, huh?" Every time he heard those words he would mutter to himself, "Why am I working every day of the year for something that isn't producing any results?"

Amano had entered the university not because he liked researching crystal growth methods. As a teenager, inspired by Masatoshi Shima who had designed the Microprocessor "Intel 4004" and the Masterpiece "Intel 8080" in the US, he had decided to pursue electrical engineering. However, at that time there were no computer specialists teaching at Nagoya University. Amano also felt that, with a few exceptions, all professors and their students were just "doing research for the sake of research".

One of the exceptions was Akasaki, who had come from Matsushita Electric when Amano was still a junior in the undergraduate program at Nagoya University. Amano felt that at least Professor Akasaki was actually trying to create something new. Amano visited Akasaki even before he officially applied to Akasaki's class and persistently asked him if he could be admitted as his understudy. His wish was finally granted in April, 1983. Amano was now completely satisfied with his research, even if it did seem like stacking pebbles up one by one like in the Styx. Even after proceeding to the graduate program, he remained in Akasaki's class.

He reluctantly submitted his master's thesis about a month before it was due, since during his two years there, he had not really accomplished anything worthwhile. He began to doubt if it was physically possible to develop the GaN crystals in the particular way he aspired to. Nonetheless, he persisted. The scariest aspect was that if it really was impossible, his research would end up as a complete waste of time. Amano could not imagine what the future held for him if he pursued a doctoral degree. At times it seemed almost like a bottomless pit ready to swallow him whole. However, there was one sample right in front of him that gave him some reassurance that his efforts were indeed not in vain. Amano, praying for the best, began tests with the X-ray diffraction meter.

Challenging untrodden realms

The story of the blue light-emitting diode (blue LED) had begun in 1963. In that year, Akasaki, thanks to Konosuke Matsushita (1894–1979), who had established the company, joined the Tokyo Research Institute of Matsushita Electric as the fourth group leader.

The period from the 1950s till the 1960s saw a boom in the electronics industry, centered on activities in these companies' research laboratories. In 1961, Matsushita Electric established Tokyo Research Institute in Kawasaki city. It was decided not to use someone from within their own company to oversee the work there, and opted instead to search around in universities for dedicated and talented overseers. Akasaki, who developed the VPE method to produce crystals for semiconductors, was selected.

Soon Akasaki began to invest efforts into developing new methods to grow the crystals for the semiconductors, using gallium phosphide (GaP) and gallium arsenide (GaAs). He chose not to use silicon (Si), which was already being used as the base material for electronics, but opted for a previously untried course, simply for the sake of innovation. The main reason for this was that he was just really anxious to conduct experiments with chemical compounds to develop the crystals.

"I also plan to try experiments in developing crystals through aluminum nitride (AlN) and GaN, which is already known to be extremely difficult," he said at the time. Akasaki became extremely determined in 1965, his second year in Matsushita. He explained, "If humans succeed in developing these crystals, then we will also be able to generate blue to ultraviolet light. Maybe it is possible to do the same things with zinc selenide (ZnSe). However, since it is much more fragile than GaN, we cannot make robust LEDs and laser diodes. I really enjoyed a good challenge. The tougher the better. I wanted to achieve results no one had previously achieved. This was the only thing I could think about at the time."

In 1969, Herbert Paul Maruska and his colleagues at the RCA Laboratories in America, succeeded in developing polycrystalline GaN growth by using the VPE method. Akasaki thought he would lose the race. In 1973, he decided to devote all his life to the research on single crystal growth of GaN.

However, researchers all over the world had given up on the single crystal growth of GaN, faced with the extreme difficulties involved. Jacques I. Pankove, who created the blue LED of GaN in 1971 at the RCA research center in America, also stopped his research in 1974. Akasaki's last rival, the research team members of Philips Laboratory in France, also decided to abandon their research in 1977. Given the circumstances, even the management of Matsushita called Akasaki and said, "We highly doubt that anything will come from further research involved with this GaN. What do you possibly intend to do with it?" The company had many other important research themes that needed an equal amount of attention.

Despite the air of futility, Akasaki refused to quit as he had noticed a sense of a "gift" implied by GaN as a material. He said, "I had sometimes found a tiny crystal within the very rough material grown (although microscopic) which I believed held great potential.

I also found at first glance, after looking at what had happened in the reaction tube after it had finally grown, that there appeared to be a "group of tiny needles which shone." After experiencing this firsthand, intuition told me of the great innate abilities which GaN held for light emitting." He felt that this tiny needle had the potential to spread all over the crystal. Therefore, he was not willing to give up at this point.

Finally, after careful consideration, he concluded that the MOVPE method, which uses organic metal compounds of gallium as a source of gallium atoms (Ga), would be the most suitable to develop GaN crystals. Then, Akasaki decided to accept an invitation to go back to Nagoya University as a professor and immediately began working on the MOVPE method. He was fifty-two years old then and still taking on new challenges.

The first paradigm disruption

Why did Akasaki try the impossible in attempting to disrupt the crystal growth paradigm? His answer: "As I had decided to use the MOVPE method, I needed to choose a substrate to grow GaN crystals which could naturally withstand temperatures of over 1000°C of heat in its natural form. In addition, I had to choose a substrate that could withstand exposure to ammonia which provides a source of nitrogen atoms (N). At first, I thought for the time being that sapphire would be the best material for the substrate. However, the sapphire crystal, Al_2O_3, has structure and bond length completely different from the GaN crystal. Thus, sapphire crystals Al_2O_3 never satisfy the law of the "lattice-matched condition" to grow GaN crystals, which is the most important rule for crystallography. I braced myself as I decided to challenge this crystal growth paradigm. If I were to successfully disrupt this paradigm and "graft a bamboo shoot on a tree," which was theoretically impossible, I would be able to grow any crystals.

"So, how could I smoothly connect a Lego block to something which is not a Lego block? The answer: I needed to insert a sponge between two different kinds of bricks. In other words, first I should pile a soft sponge as a buffer layer on top of the sapphire. Then, I would need to grow hard GaN crystals on top of the soft sponge." Masahiro Akiyama from Oki Electric had already been successful with this "buffer method" before.

Akasaki wrote down four possible component materials for the buffer: AlN, GaN, silicon carbide (SiC), and zinc oxide (ZnO). He was already quite familiar with the first from experiments conducted since 1965.

Then, in 1985, Amano decided to take on the challenge that seemed almost utterly hopeless. He described the feeling as "clutching at a straw".

Amano stated, "I was really disappointed when I completed my master's thesis, given my dirty looking GaN grown on sapphire, because I knew that it was a complete failure. Still, I continued to try the experiment after finishing my thesis. One day, the condition of the furnace for the experiment was not up to par and it just did not seem to get as hot as I needed it to. The temperature required for GaN growth is around 1000°C, and for AlN growth it is around 1200°C. Therefore, the temperature difference for the growing processes was determined to be too great given the required temperatures. Then, it dawned on me to try to use AlN as a buffer on that same day when the furnace was not working properly. I already knew that AlN has a higher density than that of GaN, given its hexagonal shape and characteristics and that AlN would be able to cover the sapphire surface more quickly. So, if one were to first use the AlN buffer, it should uniformly cover the sapphire. Then, after it uniformly covers the sapphire, GaN should be able to grow properly."

It took only a few hours before the moment when the hard labor bore fruit. He succeeded in creating a thin film of GaN on his first attempt that day. By the result shown by measuring the sample with X-rays, he proved that it certainly was a high quality singular crystal. It was at that moment that a perfect GaN single crystal was created for the first time in the world.

The second paradigm disruption

After Akasaki and Amano had accomplished this first paradigm disruption in February 1985, there still seemed to be no researchers who had their sights focused on GaN. This accomplishment marked the starting point for these two men.

After they succeeded in creating almost perfect GaN semiconductors, their next goal was to realize "p-type" GaN semiconductors. Once again, they faced a potentially impossible task. A theoretical physicist claimed that if there was even just a slight amount of nitrogen vacancy it could become a strong "n-type." If so, the p-type was theoretically impossible to create.

Akasaki and Amano tried to add zinc (Zn) to the process but to no avail. Was the theorist right? Was it impossible to realize the p-type GaN? If that held true, then it certainly would be impossible to develop a blue LED that could be used practically.

Amano was now entering his second year in the doctoral program and he took that as an opportunity to join the NTT (Nippon Telegraph and Telephone Corporation) Musashino Research and Development Center as a trainee for two months during his summer vacation. During his last week, he was given permission to use an electron microscope from NTT. With great concentration, he carefully observed the GaN sample that he had previously added the Zn into.

Amano noticed something odd when he observed the sample through the electron microscope. When GaN became exposed to the electron beam, the blue light emissions emerging from the zinc became brighter. He immediately phoned Akasaki and reported his observation. Akasaki thought that the Zn might indeed be activated by the electron beam.

In other words, the reason it couldn't become the desired p-type GaN was because the hydrogen atoms end up connecting themselves to the Zn atoms. However, in the areas that were exposed to the electron beam, the electronic state was transformed. It became possible for light atoms like hydrogen to be taken. If hydrogen could be taken, it would activate "dead" Zn atoms. In short, when exposed to the electron beam, the blue emission shines brighter, which shows the possibility of becoming a p-type GaN! Akasaki directed Amano to continue experiments in order to gather more information.

Akasaki waited anxiously for Amano to bring back all the data he could possibly get from NTT, to Nagoya. Then they took the sample to Toyota Gosei, which had the latest electron microscope available, and re-examined the sample under the microscope.

The results corroborated the discovery at NTT. They put the sample, which had successfully grown the GaN through adding zinc to the AlN buffer, into the electron microscope. After thoroughly using a vacuum, they once again exposed the sample to the electron beam. Observing the experiment through the monitor and through the reflection, they saw that the blue emission gradually grew brighter and brighter.

In the experiment, the "dead" zinc atoms were activated when exposed to the electron beam. Though it was still not the p-type GaN, it was something that provided insight into the p-type GaN. They thought that maybe it would be more effective to use magnesium (Mg), which was easier to ionize and move than Zn, to create a "p-type transmission" given that the electron state of the sample changed and the Zn was activated.

As a next step, Akasaki and Amano replaced the Zn impurities with Mg ones. They ordered an organic Mg metal compound from an American company. A year later, at the end of 1988, they finally received the material and immediately began to experiment with it.

Their expectation had proved to be exactly correct. They exposed the GaN crystal with Mg, to the electron beam. They noticed that the blue emissions proved to be amazingly brighter than before. Then, Akasaki and Amano measured the "Hall effect." As a result, they found that the electron holes carried currents. The moment that Akasaki and Amano had been waiting for had finally arrived. They had obtained and created the p-type GaN and had the proof right before them. It was January 1989.

They announced their success of the bright blue emitting diode, which was harnessed by a transparent GaN, at the American Electrochemical Society in Los Angeles, in May of that same year, 1989. At the same conference, Martin Fleischmann and Stanley Pons announced their discoveries in cold fusion for the first time. However, it was Akasaki's presentation that most impressed the audience and drew endless applause.

The third paradigm disruption

The processes used until the successful creation of the blue LED have been previously noted in the first two paradigm disruptions. Furthermore, the third paradigm disruption is related as well. It was the successful result of combining GaN with indium nitride (InN), which enabled the emission of blue light. In 1989, Takashi Matsuoka (1953–present, Photo 3) of NTT Optoelectronics Research Laboratories, and his research members, were the ones who accomplished that.

Photo 3 Takashi Matsuoka (taken at NTT Basic Research Laboratory, Atsugi, on March 25, 2001)

Matsuoka completed his master's program and joined Electrical Communication Laboratory at Nippon Telegraph and Telephone Public Corporation in 1978. In 1980, he began to develop the semiconductor laser for optical communications.

During the latter part of the 1970s, the management of Nippon Telegraph and Telephone Public Corporation (NTT) engaged in a national project to create an optical fiber network that would span all over Japan. However, the most important semiconductor laser to emit the light was not yet ready. Therefore, they needed the device to be up and running. It was the top priority for NTT, as well as for the researchers in the laboratories.

Two years later, Matsuoka succeeded in completing the infrared semiconductor laser needed for optical fiber communication.

In 1985, after Matsuoka had finished developing the laser, he was given some time to consider the next research theme. By now, other scientists had succeeded in emitting light in red. Matsuoka thought over some challenging experiments to create the emission of green or blue light. In order to do that, he pondered over whether he should use ZnSe or GaN.

In Matsuoka's hypothesis, theoretically, GaN was a better substance for semiconductor lasers because it could be altered in large-scale proportions with regard to the color of the light it emitted. In addition, no one had previously succeeded in developing a combined crystal consisting of indium instead of gallium. He thought it a worthy challenge.

At that point, Matsuoka offered to transfer himself to Ibaraki Research Laboratory where researchers had already grown crystals of GaN. He had learned the process from a recruit named Toru Sasaki.

Sasaki talked about his manager, Matsuoka. "At the time, I desired to know more about things unknown as opposed to lending my assistance to developing some fine devices. Then, Matsuoka came to our laboratory. His aim was to develop the blue semiconductor laser. With his arrival, it became a priority to help him reach his goal, not merely to enjoy physics. So, I was introduced to Akasaki's buffer method. We did our best to find out as much as Akasaki himself understood through experimenting and changing the growth conditions for about an entire year."

Matsuoka, who caught up with Akasaki's research in a year, worked on his original research subject, indium. However, all his experiments ended in failure. When he put the organic indium gas into the furnace, the only thing he could get was soot. Finally, he overcame this problem by using nitrogen gas as a "carrier gas," instead of hydrogen gas. This idea was, actually, quite ingenious.

The crystal growth MOVPE method which commonly used hydrogen gas would stop functioning if impurities got mixed up with the carrier gas. The quality of the device to purify hydrogen was a lot better than the devices used to purify other gases. However, Matsuoka thought that it was certainly possible that the hydrogen gas was a cause of the failure. He re-considered the physics to understand how to better conduct the experiment and discovered that it would be more feasible to use nitrogen as a carrier gas to grow InN, because nitrogen can enhance the resolution of ammonia raw material. Matsuoka decided to switch from hydrogen gas to nitrogen gas.

In the first experiment with nitrogen gas in 1988, Matsuoka saw some surprising results. Though only soot was generated under the growth condition with hydrogen gas, he did in fact succeed in developing a wonderful InN sample.

Next, he had to successfully combine InN and GaN. He continued his experiments by changing the quantity of the gas each time, but unfortunately it did not act. One day when he was almost ready to give up, he decided to return to the basics. Indium is very different from gallium-based gases with their ability to unite with nitrogen.

Matsuoka continued to calculate the chemical reactions of gallium and indium. He discovered that it was possible to make the alloy if one were to add sixteen thousand times nitrogen ammonia at 800°C.

Thus, Matsuoka and his research team succeeded in developing the combined alloy form of InN and GaN in March 1989. Through this discovery, the last hurdle was overcome and their results were set as a precedent for future experiments to introduce indium.

2. Serendipity

Shuji Nakamura's disgrace

In the summer of 1989, Matsuoka and his research members announced at an international conference that they had reached the third paradigm disruption. At the same time, Shuji Nakamura (1954–present, Photo 4) of Nichia, began the experiment by using the MOVPE device, which was set up by his research team members while he was in the US.

Photo 4　Shuji Nakamura (taken at University of California, Santa Barbara, on December 12, 2000)

Nakamura, who had just returned from the University of Florida, remembered his disappointing experiences during his study abroad in America. He was not admitted as a researcher. Besides, he was already in his mid-thirties and still had much to accomplish. His passionate ambition was to become a physicist and to live in a manner in which he could be true to himself. To do that, he needed to produce some respectable achievements—and as quickly as possible.

Nakamura's life turning moments had always emerged from utter failure.

Nakamura was born in Sata Cape in Shikoku in May 1954. He originally wanted to become a physicist because he had admired a cartoon character named Dr. Ochanomizu in "Astro Boy" during his childhood. He longed to study the beautiful laws upon which the universe was founded. However, his high school teacher had warned him, "If you major in physics at university, it would be impossible to get a job." Thus, he decided to major in electronic engineering at university. He chose the subject because he thought "electronic" was closely related to physics.

This decision led to his first disappointment. He soon discovered that there were no professors in his department who could quench his academic curiosity and thirst for knowledge. He ended up shutting himself away in his dormitory room and reading any and all physics books he could get his hands on. He remembers, "Aside from the material by Shinichiro Tomonaga, I could not really understand what any of the stuff I was reading meant. I concluded that the translators did not properly understand what they were translating, so I decided to find and read the original versions. They were certainly much easier to understand.

"However, I did not learn directly from any good professors. Even today I still feel inadequate in properly understanding some theories in physics. I longed to know the secrets of the universe. My biggest regret, even today, is that I did not choose to major in physics."

The second major regret Nakamura experienced came just after his first appointment as a researcher. He completed his master's program in 1979 and joined Nichia in Anan, Tokushima. Nichia, a venture business, was founded in 1956 as a fluorescent substance maker. Through popularizing the fluorescent light, it has grown into a flourishing business.

The president of Nichia, Nobuo Ogawa (1912–2002), had been sent to the battlefront of Guadalcanal as a pharmacist during the Second World War. There he experienced death and tragedy all around him before returning home. The president understood Nakamura's immense desire and placed him in the development department.

Short of funds, Nakamura soon discovered that it was not that simple to manufacture semiconductors and compete with the other companies. Besides, he was young and self-taught and had still much to learn before he could compete in the field. The Nichia management felt that their president had been reckless in his decision and often mocked Nakamura because he was so intent on trying to become a researcher. They felt that he didn't quite know his place.

Nakamura remarks, "After I joined Nichia, I felt that my life had really come to an end. I needed to develop devices for practical use; otherwise I would be seen as valueless in the eyes of the company. Several months after that, I stopped reading books about physics. I decided to abandon theory and concentrate on developing devices."

After Nakamura joined Nichia, he developed polycrystals consisting of GaP by following the example of others and sold a considerable number of them by himself, which contributed to Nichia's earnings as well. But it took him 10 years to conclude that his efforts were not as fruitful as they should have been. He had been developing devices that lacked originality, and furthermore could be easily developed by any of the larger manufacturers. Without originality, it would be difficult to succeed in the market. At the same time, Nakamura was plagued by the feeling that he was merely going through the motions and had been unable to achieve anything truly significant.

He sums up these feelings thus: "Up until then, I had researched, developed, and promoted semiconductor devices alone and made several tens of thousands US dollars of profit for Nichia over that ten-year period. I was the only one in the company who had developed fluorescent substances and who had come up with anything new and original. The president of Nichia also knew how difficult it was to develop new things. He also knew that developing an already overly produced item and selling it in a highly saturated market was not the best approach."

Nakamura urged his president, Ogawa, to work on developing a blue LED in order to capitalize on the first mover approach. He thought that a small business, like Nichia, needed to create a new market or it would just end up being trampled over by all the other companies. Ogawa, who was very experienced, perceived the situation a little differently, concerned that they might end up getting stuck in a disadvantageous situation if they were to open up a new market.

Ogawa's experience with fluorescent light actually had its origins in his days as a prisoner of war after being captured by the US army. The first time he saw fluorescent light was after he was taken to Guadalcanal. Ogawa wanted to recreate that light which he saw, and had consequently founded Nichia. He also understood Nakamura's youthful exuberance and in the end, gave in to his persistence. Ogawa began with an investment of 300 million yen (approximately 3 million US dollars) in the project.

But unbeknownst to Nakamura, he was about to experience even more disappointment during his time spent studying abroad in the United States in 1988.

Nakamura joined a research team at the University of Florida in March of that year. He chose the University of Florida because an upperclassman at his university named Shiro Sakai had already begun research on growing crystals. Nakamura, as mentioned earlier, did not have a physics degree, nor had he written an academic paper on physics.

On his time as a member of the research team at the University of Florida, Nakamura says, "I experienced a great deal of disgrace there. None of the other researchers acknowledged me as their fellow equal. They regarded my life's work as nothing deserving of respect. While they did not invite me to their conferences or research meetings, they always called me to repair something. Although I had heard that the University of Florida already had a MOVPE device, it was still incomplete. So, I took it upon myself to construct the device and in the end I did finish the MOVPE with some graduate students from Korea and China. It took us about a year and we did the final touches on it just about a month before my last day at the University of Florida."

Shuji Nakamura's stroke of luck

Nakamura returned to Nichia of Japan in April 1989. At that pivotal time, he discovered hidden reserves of determination within himself. He vowed to forge ahead regardless of what anyone said—even his superiors.

The only person he found he could trust was Sakai, who had already obtained a position at the University of Tokushima. After that, Nakamura remodeled Nichia's MOVPE device to fit the needs of his specific crystal growth according to what he had learned from Sakai.

One year later, in 1990, Shuji Nakamura struck gold, albeit unexpectedly. He discovered the ideal conditions suitable for growing high quality single GaN crystals. This run-in with serendipity came from an original two-flow method. He named it himself, in fact. The method is to place source gas into a furnace from the side and then add a large quantity of a mixture of nitrogen and hydrogen gas from the top.

Nakamura describes the process, "For one year I had tried over and over again to grow the crystal after I got the crystal growth apparatus. However, I just could not seem to create the thin film of GaN I had been trying so hard to make. Because when you think about it, the required growth temperature is a naturally high temperature over 1000°C. When the source gas warms up it naturally ends up rising to the top due to convection even when it is put in from the side. This made me think that maybe the source gas would not work well on the sapphire surface. Then it dawned on me to try to put both of the gases in from the top to pin down the source gas to eliminate the convection problem."

Nakamura succeeded with this two-flow method on his first try. He created a flawless, thin film of GaN on top of the sapphire. The light emissions from this were even more overwhelming than those produced by the forefront researchers like Akasaki and Amano. He used the buffer technology designed by Akasaki and Amano in the process as well. Six months later, he decided to try to use GaN instead of AlN as a buffer layer. As a result, Nakamura achieved even higher performance than ever before, higher even than those reported by Akasaki and Amano themselves a few years earlier. Even after that he continued to get amazingly good characteristics in every experiment he did. Nakamura immediately applied for a patent for the GaN buffer technology.

In the patent system of Japan, it awards a patent to the first person who applies as the official inventor, not the first person showing evidence that he has successfully invented the item. (However, if it has already been published as an academic paper or is something well known. it cannot be patented.) Therefore, even though Akasaki had already documented his success "using GaN as a buffer" several years earlier, it did not interfere with Nakamura's application because Akasaki never applied for a patent. Actually, Takao Nagatomo, a professor at Shibaura Institute of Technology, as well as Matsuoka and his team of researchers, had also already successfully used GaN as a buffer and had announced their findings at conferences. However, since there was no formal documentation, there was nothing to stop Nakamura from patenting it. Nichia was now recognized as the holder for the patent of the GaN buffer making system, making it their second hugely successful value-adding patent. When Nakamura developed a p-type GaN, he established his originality.

Despite this success, the question remained, why were they unable to activate magnesium atoms? Naruhito Iwasa, who worked under Nakamura, found something after continued research on the experiment.

The resistance value is changed when ammonia is poured as a source gas over a magnesium–GaN mixture and then heated. Iwasa guessed that after the ammonia broke down, the remaining, undamaged hydrogen had something wrong with it. Hydrogen should completely stick to magnesium without moving around at all.

So, Iwasa removed the hydrogen that had been concealed inside of the crystal by simply treating it with heat, thus allowing the GaN to become the desired p-type. It was December 1991. It became the strongest patent in history acquired by Nichia.

Nakamura not only mastered existing methods discussed in the first and second paradigm disruptions, but also created his own methods. However, he did in fact have a hard time with the introduction of the third paradigm disruption. Actually, having already written a paper in just one month, he suddenly stopped writing his academic paper. He didn't write anything else after that for six months, until July 1992.

During that time, Nakamura consulted Matsuoka on the method to be used to introduce indium, who advised him to follow the steps he had outlined in his paper. Nakamura followed Matsuoka's recipe step by step, except that he used his own two-flow method. As a result, the quality of alloy was overwhelmingly better than the alloy results indicated in Matsuoka's paper.

Nakamura succeeded in putting the blue LED diode into practical use in 1992. He did not accomplish this feat in isolation. This historical achievement came from the "integration" of experiences from each of the paradigm disruptions.

On a side note, Matsuoka and his research members suddenly and unexpectedly disappeared from the history of this innovation, not to be heard of again. Mystery still surrounds their disappearance from their promising careers.

3. What were NTT and the other big enterprises doing?

"Do you know what happened to Takashi Matsuoka and his old research members at NTT?"

I posed this question to Shuji Nakamura himself. I could tell by his countenance that he had no idea, and then he tried to change the topic.

"If I had not joined Nichia, probably I would not have succeeded in the research of GaN. At that time, Nichia was still a small company, so I could ignore what everyone was asking me to do and started developing devices I thought would actually be of use. Today, Nichia is one of the biggest enterprises in the semiconductor industry of Japan, and if I were to attempt to do the same thing I did before, I would surely have been fired. Now, all the rules are clearly written on paper."

Like Nakamura, Matsuoka and his team members were researchers in an enterprise, so he had opportunities to lead the innovation. However, Matsuoka and the names of his research members suddenly disappeared from the history of innovation when Nakamura published his academic paper on the blue LED in 1992.

Why did NTT quit the research?

Why did Matsuoka and his research members suddenly disappear? The reason behind this is that the NTT management ordered them to abandon the research.

But why did the management stop the research, especially at such a critical juncture for innovation?

The answer is not for some trivial reason like the management wanting to avoid risks. As most of the management members were excellent scientists, who had successfully led several research projects themselves, there must have been some other practical and fundamental reasons behind their departure.

In 1987, Matsuoka, who was obsessed with creating blue LED, transferred to Ibaraki Research Laboratory. In the same year, the NTT management decided to move another research team working on ZnSe to the Ibaraki Research Laboratory. Thus, there was now the research team working with ZnSe alongside the team working with GaN. In the eyes of the management, this was a logical pairing since both teams were ostensibly working toward the same goal: the realization of blue light emission. Unfortunately, that pairing was just the beginning of an unpleasant series of events for both parties.

In those days, the NTT management provided their researchers with enough funds in the same way as Bell Telephone Laboratory did. Even with regard to research themes, the management respected their aims. They wished to make sure that the researchers did not fall behind the others at the forefront of technology by neglecting basic research. In Japan, as most researchers in universities seldom enjoyed working on projects related to innovation, the NTT management would have felt a sense of pride since these accomplishments had carried them to the forefront of technology.

However, the situation changed after 1990. Competition became fierce, posing a threat to the researchers' freedom on research projects. They were forced to make tough decisions in research courses and in sticking steadfastly to them.

Such discussions were heard over and over again at the headquarters of the research and development department. Management would need to evaluate each research laboratory according to its value. As if those decisions were not taxing enough, there were imminent time constraints. Then, someone brought up a new topic: "Do we really need to focus so much on research here at NTT? Why don't we just purchase it from an outside source?"

There was a meeting at the headquarters to discuss the development of a strategy for the future. It was decided that the company needed to focus all their funds and resources on the most important things at hand and to shut down the research projects that were not directly related to their most profitable branches. The project for the blue LED, as it was unrelated to information and communications, was unfortunately dissolved.

In this situation, it was impossible for any group to continue with more than one research project at a time. Consequently, the management now had to make the choice of either pursuing research related to ZnSe or to GaN.

Which of the two was more important?

The method they used to come to a decision was quite simply to look at the factors of relevancy and how intertwined they were to their new strategy.

The first factor was the likelihood of successfully creating the p-type semiconductor. The ZnSe had already proved itself successful in 1990. In addition, 3M in the United States announced their success in pulse laser oscillation at room temperature using ZnSe to emit blue light, at the American Device Research Conference, which led to quite a sensation at the time, in June 1991.

On the other hand, GaN had had no success in developing the p-type, except for Akasaki's team which had publicized a success in developing the p-type through electron beam exposure in several academic conferences and journals. Matsuoka and his team members had succeeded in adding indium to GaN, but the hard part was in developing a p-type device with this technology. Therefore, it was almost impossible to nominate the research related to GaN over the research related to ZnSe.

Still, the NTT management could not just order Matsuoka and his team to stop all research on GaN and concentrate fully on the research of ZnSe. After all, as a public research institution, it had a tradition of respecting the ambition of researchers. Sasaki revealed, "Actually, I received a challenge from the head of the laboratory that if we could succeed in creating the blue light semiconductor laser using ZnSe by the following March, we would be able to continue further research on GaN."

In response, Matsuoka said, "We have to make our device light up, no matter what. Even though the head says he doesn't care which material we use, it's not practical to think that we can make GaN shine by March. But, if we do succeed he will let us pursue any research of our choosing. So, let's first focus on getting ZnSe to shine and then we can focus later on trying it with GaN again."

Confronted with this challenge, Matsuoka and his team worked frantically. On 17th of March 1992, they succeeded in creating the pulse oscillation semiconductor laser using ZnSe. It was certainly a splendid achievement.

However, the management was averse to the idea of supporting projects that did not directly relate to their current business focus and further diverted money away from that focus. Thus, the GaN project was canceled on the last day of March.

Within a year after that, the news that Nichia had succeeded in developing the blue LED spread all over the world. Moreover, the researchers at Nichia held a demonstration of the blue LED with GaN. The NTT research management was quite surprised by how brightly it shone. However, since the management members had already endorsed ZnSe, it would have been virtually impossible to get them to retract their decision and allow Matsuoka's team to reopen the project.

Takashi Matsuoka continued to appeal to management to understand the importance behind the research of GaN and allow them to restart it. The NTT management did in fact end up allowing Matsuoka to resume his research two years later because everyone had to admit by 1994 that the blue LED created by Nichia was very impressive and management thought that this might be the last worthwhile innovation of the twentieth century. However, Matsuoka had fallen way behind due to troubles with finding enough team members and lack of sufficient funding. Ultimately, these difficulties set them back by about two years.

Sasaki, on the other hand, had been transferred to a subsidiary company, NTT Electronics, where he began working on producing a semiconductor laser for optical communications.

I asked him, "Why did you just give up after being ordered to stop the research on GaN?"

Sasaki answered, "Which one do you think should have been chosen: ZnSe or GaN? No one could really insist on choosing to give up the research of ZnSe while continuing with the research on GaN. Actually, deep down I too chose to insist on quitting ZnSe and picking GaN because I felt that GaN was quite simply overwhelmingly better than ZnSe. For example, with GaN you can easily make and measure holes using an electrode, even though that's only a small advantage. In contrast, doing the same thing with ZnSe is quite difficult. Sticking by this point, I felt that GaN was the better option given the quality of the surface. Tallying up the pros and cons, I just felt that GaN, which, though it looked like nothing more than a crude rock, had the potential for greatness."

His words were very important in analyzing innovation management. He felt strongly that the ZnSe laser would break pretty easily, whereas if the laser were to possess GaN, it would also bring with it stronger durability. This substance certainly had the potential for greatness. But he didn't realize exactly what it was until about a year later—after he had already stopped research on GaN.

Toshiba case

By coincidence, Yasuo Ohba, who first began the research on GaN at the Toshiba Research and Development Center, echoed what Sasaki had said.

"Having done research on very small flaws during my student days, I had always felt that ZnSe's physics characteristics were rather poor. The failure of ZnSe is easier to find because ZnSe is a weak material. If I try to develop something with such a weak material, I do not think I will be able to succeed," Ohba said.

In 1983, the Toshiba Synthesis Research Laboratory hired Ohba after he completed his doctoral program. He was then placed in the electric parts laboratory and did research on crystal growth for the red laser diode materials that would eventually be used in DVD technology.

In 1985, an international conference was held. At the conference, coincidentally, Toshiba, NEC, and Sony all announced at the same time their success in the creation of a red laser diode with continuous wave oscillation sustained at room temperature. Ohba had the opportunity to be the announcer for Toshiba. He had replaced the pointer of the helium neon laser (0.633 micrometer wave length), which the conference members prepared, with semiconductor laser (0.67 micrometer wave length), which he had developed himself. At first, the audience couldn't seem to recognize what was different, but shortly after there was applause, starting from the front row as people started to notice, until finally the whole audience joined in.

In 1988, Ohba finished his work with the red laser diode for DVDs and then began to search for a new target. The next obvious choice was the creation of green light emissions and blue light emissions, since no one had succeeded in accomplishing that as yet.

If this direction were to be taken, then which material should be used for this challenge? ZnSe or GaN?

Ohba did not hesitate to choose GaN. His motivation is apparent in his statement: "Given my gut feeling which told me not to use ZnSe, I must say that I tried to take on challenges no one else had conquered yet. I let my motivation soar sky high as I chose to do something no one else had done before, which was to create blue light emissions. So, I thought about how I should approach the new challenge at hand. I chose GaN for the material. Then my team and I worked alongside the research team that was researching the ZnSe. We managed to develop a new MOVPE device within a year and began to do research on how to grow crystals on GaN by ourselves in 1990."

By that time, Akasaki had already announced his buffer technology findings. However, Ohba decided that he did not want to just imitate what other people had already done, like Sasaki had in his younger days.

After Nakamura reported his invention of the two-flow method in 1991, he announced his development of the blue LED (which had amazingly high quality) in the following year. Still, Ohba stubbornly continued to try growing GaN using his own methods.

Here is an interesting fact. Toshiba management announced that its researchers had succeeded in developing the blue light semiconductor pulse laser, based on GaN at room temperature, during a press conference on September 11, 1996. It was the second achievement in the world, putting them right behind Nichia. However, Ohba was not one of the members on the research team who had achieved it. Kazuhiko Itaya, the team leader for this particular crystal growth project, was introduced as a new young talent who had a bright future ahead of him. It was his name that appeared in the limelight of the article chronicling the development of the blue light emitting laser, in a section called "Technological Gurus," on September 30, in the "Japan Economy Newspaper." Also listed in the "Technological Gurus" column were statements about Makoto Azuma (51), who was the head of the material and device laboratory at the development and research center. He had been at the vanguard in the development of the blue light emitting laser. Under Azuma, Masaru Nakamura (45), the laboratory leader and Kazuhiko Itaya (37), the lead researchers, who were both veterans, were joined by new young researchers at the forefront of the research, on September 30.

The management of Toshiba had a strong desire, dating back to the 1980s, to develop LED and laser technologies. So did the researchers who had dedicated a great deal of time and research on crystal growth with ZnSe in order to realize blue LED. This organization's motivational drive came from the self-confident, forward-looking Toshiba, which was then leading the semiconductor device industry in Japan.

The Toshiba management immediately decided to change the research focus from ZnSe to GaN. The implementation was lightning quick with the management allowing Ohba to continue his research on GaN using his own methods, and the president of Toshiba ordering the other researchers to continue doing research on ZnSe. Then, Azuma, the research director for both projects, received detailed performance reports.

Initially, the new team members had a difficult time after they announced their success with the GaN blue light semiconductor laser to the press. Though the management of Toshiba strongly hoped that the researchers would be able to realize blue light emitting continuous wave oscillation at room temperature as quickly as possible, it certainly took them quite a long time before they had any verifiable success. For three years it appeared that their research was stymied. During that period, Cree Inc. of the United States, Fujitsu, and some other enterprises announced their success one after another. What was going on at Toshiba that brought about this misfortune?

Shingi slipped out of our hands

As chronicled in the previous section, the researchers at Toshiba, were the second team in the world to succeed in creating the GaN semiconductor laser pulse oscillation at room temperature. Then, they made absolutely no progress for the next three years. They also could not seem to create continuous wave oscillation, either. Why?

As the researchers had perfected the lightning fast pulse oscillation, did it perhaps have to do with them misunderstanding the paradigm disruption inherent in their own innovation?

The leader of the research team made a plan to analyze, in great detail, the factors that made up the technology of the semiconductor laser. The first step was to maximize the amplification by effectively confining the light. That part could not have turned out any better than it did. The second part to analyze was the technology of the process for developing the device structure. The team accomplished that as well, without damaging the semiconductor. The third aspect was the shape of the electrodes. The research team calculated how best to make the shape in a way to avoid overheating due to running so much current through the device.

This kind of method was the model in developing new devices. However, the efficacy of this kind of project management is limited to circumstances where the innovation is simply put together using existing knowledge.

In the case of the blue light semiconductor laser, the cause behind not being able to continuously operate was actually due to how perfectly the GaN single crystal was achieved. To solve this problem, the researchers needed to step back and analyze it again from a solid state physics point of view.

The researchers came to the conclusion that even if they went years back and reanalyzed it again, starting from the basics, it would be fine as long as the end result produced a viable solution. However, any researcher would be afraid to fail and would have to take responsibility for the failure, given the mounting pressures from the top management. Due to this fear, no one in the research team really had the courage to suggest returning to the basics once more, which ultimately led to Toshiba falling behind.

Kazuhiko Itaya likened the situation to Eyo, a strange festival of the Bizen Saidaiji-temple in Japan.

"At this festival," he said, "men scramble for a Shingi, a holy tree. Some men have been known to win the contest twice or more during their lifetimes. The blue light laser is this Shingi for us. It slipped out of our hands, though we had the opportunity to touch it."

However, the researchers at Toshiba did not take the time to consider the scientific reasons on why the "Shingi" had slipped out from under them. As the researchers watched the shadow of what they were looking for fade away more and more, anxiety set in.

The Sony case

Why did the other big companies fall behind as well? Simply put, the ultimate decision by management to choose ZnSe instead of GaN. The biggest problem was that management had continued to ignore GaN for several years, even after the Nichia management had announced to the press that they had begun mass production of their blue LED in 1993. I'll explore the case of Sony to see if we can unearth why they decided to hold fast with ZnSe.

Sony goes by its motto "frontiers for innovation." Its management had high aspirations in the future development of blue light emissions. In fact, they were even more obsessed than Toshiba was. This turned out to be, ironically enough, the primary cause behind their lack of success. Only in 1997 did the Sony management finally decide to give up trying to develop the blue LED with ZnSe.

They first began to develop the blue LED with ZnSe back in the 1980s. While other enterprises were withdrawing one by one on the research of ZnSe, Sony continued to invest extensive resources and time in more than thirty researchers trying to get ZnSe to work. They finally succeeded in creating a blue light laser with the ZnSe. In 1997, it was able to sustain continuous wave oscillation for three hundred hours. However, Nichia announced that its blue light laser with GaN was able to sustain continuous oscillation for over a thousand hours. Sony had no choice but to accept the fact that GaN was superior to ZnSe and so decided to adjust their tactics.

After closing operations on ZnSe, the research team was dissolved and the members were moved around to either the research team working on GaN, software, Organic EL, or any other projects that needed extra help.

The research members in the laboratory who had developed the blue light emissions with GaN caught up on the technology forefront in just half a year, and they continued to make overwhelmingly fast progress following that as well. This was because Hiroji Kawai had done research on the MOVPE method for crystal growth with regard to GaN since 1994. Kawai secretly began this research with a few other researchers as a measure of risk management. So, it was because of him and his team that, in 1997, Sony finally succeeded in developing the blue light semiconductor laser which could perform pulse oscillation at room temperature. After hearing the shocking announcement from Nichia, Sony had already completed the basis to shift their research toward GaN.

Sony led in the standardization of the Blu-ray Disk (BD) as the next generation of DVDs immediately after switching their research focus to GaN. The Sony management, at the end of 2002, announced that they had decided to share patent rights with Nichia and that they would immediately start developing and begin manufacturing these new BDs. On April 10, 2003, Sony released a video recorder which was compatible with the BDs. The memory capacity was five times that of a DVD. It could record twelve hours of movie time on a single disk. The key to their success was not only their tactful alliance with Nichia, but also the pool of research that had been accumulating with regard to crystal growth for more than a quarter century.

The NEC case

The researchers at NEC began their research on GaN in 1996, two years before Sony. However, they did not have any success with it at all in the first three years.

Of course, several researchers became irritated with the situation because there were so many excellent researchers working at NEC and they knew that they had to begin research on GaN, as soon as possible.

"However, none of them had the authority to give it the green light," Masashi Mizuta (head of the Light and Radio Devices Laboratory at NEC in 2002) said. "It is without a doubt the destiny of big enterprises. Actually, the leader of the ZnSe research team strongly hoped to do research on GaN at the same time. However, as the management at NEC continuously invested large portions of their budget in research relating to ZnSe, this made it difficult to divert those funds being invested to do research on GaN. Even though we did end up finally receiving permission to use those funds for the research, it took a long time before we got them."

Mizuta placed two of the crystal growth engineers in the GaN research team at the beginning of 1996. However, the resources they were allotted were scarce in comparison to those being used by the researchers working on ZnSe.

It was only after Nichia's success in 1997 with a blue-light laser device that worked for more than 1,000 hours that NEC management recognized the importance of GaN and decided to stop all research on ZnSe. On that topic, Mizuta continues, "The researchers at NEC had just started to begin work on developing the GaN laser by the time the researchers at Nichia had almost finished it. All we could do was follow them."

A breakthrough that would further NEC's cause came at the end of 1996, when Akira Usui of NEC's Light and Radio Devices Laboratory invented the land mark technology called the Epitaxial Lateral Overgrowth (ELO) method. This technology reduced the dislocation problem of the GaN growth dramatically. Like the way one arranges a Bonsai tree by tactfully using wires, placing a splint on the semiconductor to lead the direction for the crystal growth aided in the growth of GaN crystals.

There is more to the ELO story. Usui reported his success in reducing the dislocation of GaN at an applied physics conference in Spring 1997. Nakamura immediately asked him about the method he used. Usui answered, "I cannot talk about it in too much detail, but I used masks." Just from that one hint, Nakamura knew exactly what he had done.

Itaya, who was also there, said, "Actually, I too realized what kind of method he used thanks to the hint he gave me. However, I felt that it was much too difficult to do it that way and so I decided not to bother with it, because we needed to develop semiconductors at low cost. Nakamura immediately called his company to reproduce the Usui's method."

Nakamura immediately adapted this technology for his work in the development of the long-life laser. At the international conference in Tokushima in the fall of 1997, Nakamura reported that this success was dependent on ELO technology as he showed his blue light semiconductor laser in the hall.

Nakamura was able to catch up with his company's technology thanks to the hint given by Usui. In so doing, he was able to create a device for practical use within six months. Surely, Nakamura must be a genius to have absorbed and assimilated the intelligence from all these great and renowned researchers.

Chapter 1

Introduction

1.1 Japanese Corporates Are No Longer Innovating

1.1.1 The Science-Based Industry of Japan Faces a Crisis

The future of scientific activities in Japan hangs in the balance.

Since the beginning of this century, Japan has begun to see a decline in activities in the fields of physics and molecular biology, which form the core of science. This is most likely due to the decrease in the number of researchers who can spearhead research.

There may be a few who would question this, wondering why then in recent years, the Japanese have won the Nobel Prize almost every year. Isn't Japan second to only the United States in the field of natural sciences since we entered the 21st century? But barring very few exceptions, these awards relate to research results from more than twenty years ago.

The 2016 Nobel Prize recipient for physiology or medicine, Yoshinori Osumi, also mentioned the lack of funding for basic sciences in Japan at the awards conference and expressed that he felt "A strong sense of impending doom for the state of science in Japan."

Science-based industries are also on the decline. Japan, which was formerly known as a "scientific nation" as well as a "technology nation" and was leading the world in these spheres, has now begun

Innovation Crisis: Successes, Pitfalls, and Solutions in Japan
Eiichi Yamaguchi
Copyright © 2019 Pan Stanford Publishing Pte. Ltd.
ISBN 978-981-4774-97-0 (Hardcover), 978-0-429-44862-1 (eBook)
www.panstanford.com

to rapidly lose its distinction as the science and technology capital of the world.

Among other things, since the turn of this century, the international competitiveness in the electronics industry, including semiconductor devices and mobile phones that was Japan's forte, has plummeted. In addition, its production has halved since reaching its peak in the year 2000. The Japanese pharmaceutical industry that was ranked at the top of the science-based industries of the 21st century, also lost out to international competition in early 2000.

This means that there are no more innovations emerging from the high-tech corporations of Japan.

Amid progressive globalization, Japanese society has stuck to traditional industrial models, without pursuing innovations and models that are more in tune with the times, thus leaving the country far behind the rest of the world. Japan has lost its ability to take risks and has failed to convert many new technologies resulting from research and development into economic value.

The crisis of science is not limited to the decline in Japan's industrial competitiveness. The severe accident that happened in TEPCO's Fukushima Daiichi Nuclear Power Plant in March 2011 at once exposed the lack of scientific consideration in the management of technological enterprises. If we probe further into what caused the accident, it points us directly toward the lack of innovation in monopolistic or oligopolistic corporations.

There are many voices saying that the reason innovation has come to a standstill in this country is because "Talented entrepreneurs are no longer around" and "Japanese lack entrepreneurial spirit as they are mostly enterprise-oriented." However, it is evident that institutional and structural factors are the root causes of this problem.

The purpose of this book is not only to explore the structural factors but also explain the process of innovation from scientific discovery, with a view to providing a practical breakthrough solution to revive this scientific powerhouse. This solution will also be a prescription for corporates and provide measures to prevent accidents where science can damage society like in the nuclear power plant mishap.

Before going into specific considerations, I would like to point out why I came to pursue the theme of "science and innovation" despite being involved in physics research. I believe that sharing experiences will provide a better context on the issues being discussed in this book.

1.1.2 The Collapse of Central Research Laboratories Triggered This Crisis

Coincidentally, when I was engaged in condensed matter physics research at the Basic Research Laboratory of NTT (Nippon Telegraph and Telephone Corporation), focusing on the development of next generation semiconductors and transistors, I happened to discover that hydrogen trapped in palladium metal exhibits an unknown exothermic reaction. This gave me the chance to continue this research at the Cote d'Azur as an invited research scientist of a research institute in France for five years beginning from the year 1993.

When I returned to Japan in 1998, I was shocked. Not only the electronics industry but even large corporations in the pharmaceutical industry were shutting down or downsizing their central research laboratories one after the other, and distinguished scientists and engineers working in these labs were being forced to relocate to factories and even to sales and administrations.

This is the so-called "end of the era of central research laboratories of large corporations." "Central research laboratory" may have different names in different organizations, but it is a major division of a corporation that is primarily engaged in scientific (pure basic) research.

The central research laboratories of Japanese companies came out with several technological innovations based on cutting-edge research in the 1980s. At that time, eighty percent of the national research expenditure came from private enterprises, and university research hardly contributed toward innovations. So, research by corporations primarily drove innovation.

However, in the latter half of the 1990s, after what happened to the private research laboratories of AT&T and IBM in the United States, Japanese corporations followed suit and decided to almost completely withdraw from research. First, Hitachi shut down its basic research laboratory for all practical purposes, and the central research laboratories of NTT, NEC, and Sony were also impacted internally.

If this situation continues, there will be no more minds to spearhead technology innovations that have been supporting Japan's industry and economy. Ten to twenty years later, there will be a tangible deterioration in Japan's science and science-based industries, and Japan will definitely be left stranded.

With this conviction, I left NTT Basic Research Laboratory and took a temporary break from my physics research, and from the year 1999 I began research and policy advice on innovation strategies as part of the 21st Century Public Policy Institute, the think-tank of the Japan Business Federation. At that time, the Chairman of the Board was Naoki Tanaka, an economic critic, and the Director was Shoichiro Toyoda, Honorary Chairman of Toyota Motor Corporation.

All the researchers were economists, and in the science department the only physicist was myself. For physicists, the field of social science is a treasure house of research materials. Why did the unique technological innovation framework developed by Japan collapse? At this point, for the first time I got to learn the method of "transcending the borders of knowledge (knowledge cross-border)" between natural science and social science.

1.2   What Can Be Done to Revive Innovation?

1.2.1   Leverage Dormant Talent

How can we save the pool of world-class scientists in large corporations who are being sidelined as being "unprofitable?"

First, in order to learn how to set up a company, I decided to challenge social convention by having my wife assume the role of chief executive officer. Most women are busy raising their children until they are past forty. However, once they have raised their children and wish to get back to their jobs, the pension system poses a challenge, due to which there are hardly any corporate willing to hire women over the age of forty-five. The only jobs that they can find are those where they cannot fully utilize their expertise and creativity.

Fortunately, since my wife is a qualified pharmacist, she was able to work part-time at a pharmacy even while raising our children. I suggested to her to take the plunge and start her own pharmacy. It is a one-of-a-kind pharmacy which people in the community consider as a family pharmacy, where recommendations on the doctor's treatment can be given appropriately.

I made full use of the data available on maps to locate the medical doctors' clinics and the population distribution data to identify the area in need of a dispensary. Then I took the decision on where to open the pharmacy and rented out a small store at that location. I then invested half of my retirement money on my wife's venture. My wife, who was initially hesitant over the pharmacy venture, went on to wear the proprietor's hat, and two years later, she took a financing loan and set up a second store.

The next step was to eventually set up a high-tech start-up company.

I interviewed more than a hundred leading world-class scientists and engineers and continuously observed scientists who had left large corporations in Japan and the United States. Then, I began to explore the process through which innovation is born.

There is, in short, only one definitive treatment to free Japan from the current crisis in science and innovation. That is to encourage all the eminent scientists and engineers who have been laid off to become innovators by setting up their own start-up companies.

There is a large group of scientists in Japan, a silent majority who will be the champions of innovation.

For example, the blue light emitting diode (LED) that was in the running for the Nobel Prize in physics in the year 2014 was a culmination of research done by three physicists, namely Isamu Akasaki, Hiroshi Amano, and Shuji Nakamura from the mid-1980s to the early 1990s. In addition, the discovery of GaN crystal growth method to the invention of the quantum mechanical device structure and the development of production technology and commercialization, are the innovations of all occurred in Japan.

Moreover, we cannot overlook the fact that this achievement can be attributed not so much to the fact that Mr. Akasaki and Mr. Amano held positions in Nagoya University as the intrinsic role played by the research laboratories of corporates such as Tokyo Research Laboratory of Panasonic, Semiconductor Research Laboratory of Oki Electric, and NTT Basic Research Laboratory, as explained in Prologue.

However, many innovators like them are being neglected in the companies, and they are almost losing the avenues to innovate. If we can tap their talent and set up a company, we would be able to bring out some remarkable innovations.

Unable to contain myself, I visited the research laboratories of corporates that were about to shut down and approached these distinguished scientists, who had been asked to stop their research by the management, with my proposal of establishing a start-up company.

Among the many high-tech start-up companies that I have founded so far, I set up one such company that primarily dealt with GaN.

In order to make power transistors that can switch high voltage and high current, in conventional silicon semiconductors, the bonds of atoms forming the crystal are weak due to which they cannot withstand high voltage. Though GaN crystals help to overcome this problem, no one in the world could make them at that point, but eventually a crystal growth method was discovered and the blue LED with GaN was invented.

In the latter half of the 1990s, Hiroji Kawai of Sony Frontier Science Laboratories, appeared in Prologue, made the transistors by GaN and succeeded in bringing out a commercial design for the first time. If this design is introduced to society and converted into a product of value, it could eliminate energy loss by eight percent or more through voltage conversion at the time of distribution. This could address more than twenty nuclear power plant bases.

However, the Sony management was only interested in computers and entertainment at that time and showed absolutely no interest in Kawai's research, saying "It is enough if you can just buy us the parts" or "all basic technology is sold at the greengrocer." In other words, the management at that time was of the opinion that all basic technology such as materials or parts should be procured externally and did not seem to think that breakthrough technology was born out of science.

I exhorted him, "Your innovation will be shot down in Sony. So, please quit Sony. It is your innovation that can save sinking Japan. Let us set up a start-up company right away."

Having thus persuaded him to leave Sony, we set up a start-up company using the remaining half of my retirement amount and his entire retirement allowance in the year 2001. This start-up company endured several ups and downs, but in the year 2014, we finally succeeded in developing the world's best GaN power transistor as a quantum–mechanical device. It was still extremely difficult to raise funds. The situation in which Kawai has had to "hold onto a tightrope and wait patiently," in his own words, has been continuing for many years.

1.2.2 Developing "a Good Eye" for Innovation

Another energy start-up company, which I participated in setting up involved developing a completely new concept in storage batteries.

Hisashi Tsukamoto, who has been researching lithium ion batteries from an early stage at a Japanese battery manufacturer's laboratory, moved to the west coast of the United States and successfully set up his start-up company because he felt that it was "not possible to take on new challenges in Japan."

It was in the early summer after the TEPCO Fukushima Daiichi nuclear accident in 2011 that Tsukamoto got in touch with me. We met again the first time after many years spent together at the dark, scorching Kyoto station that was conducting high power conservation initiatives in favor of TEPCO. He told me, "I never thought that Japan was such a vulnerable country. It is a pity that there are absolutely no innovations in the power industry." He added, "Would you be willing to help me set up an energy company in Japan?"

We then launched a start-up company in a month's time. While conventional lead batteries are safe, they have a short lifespan, and though lithium ion batteries are much more efficient, the drawback in them is that they can ignite if overcharged. Our company introduced a pack containing two batteries and created a revolutionary electricity storage system called the "bind battery" in which the batteries compensate for each other's drawbacks, and not only increase the lead battery life by four times, but also act as safety valves to ensure that the lithium ion batteries never ignite. There was also the unexpected discovery that it worked perfectly well when the temperature is under forty degrees of Celsius below zero.

Also, the company succeeded in developing the "shuttle battery" that supplies hydrogen using iron powder in the fuel cell. This battery can run a car for more than two hundred kilometers using iron powder corresponding in amount to just around a dozen cola cans even in the absence of an external supply of hydrogen. Also, if iron oxide can be reduced to iron using midnight power, the iron powder can be recycled. The company treaded on thin ice while raising funds and managed to continue its development without falling from the tightrope.

As you can see from the case I was involved in, even in industries that are considered to be no longer innovating, there are many dormant seeds of innovation. The key lies in whether you have an eye to discover these potential ideas and a strong sense of determination to take them forward.

Nonetheless, in the process of setting up a start-up company, I have been confronted with situations that have often left me dumbfounded.

There is a strange perception in Japanese corporate culture that "start-up companies cannot produce breakthrough technology." Therefore, as mentioned, even if the development is complete, and only the production technology needs to be put in place, you are faced with a situation where you have hardly any investors around to fund your project.

Even amongst start-up companies, the so-called IT corporations such as in software development, do not require such a large investment. But breakthrough technology based on creation and design of new materials requires huge capital investment, and you cannot develop new industries in the absence of such funding. No matter how superior or futuristic a technology is, it is way too difficult for scientist entrepreneurs to succeed, and it is a fact that they cannot get past even the first stage.

What I realized after starting several businesses is that the number of people who have the capability to understand the ground design of innovation and conceive an idea for the future is critically low in Japan. I believe that it is absolutely essential to nurture people, whom I call "innovation sommeliers', to develop innovations for a better and brighter future.

1.2.3   Science Literacy and Trans-science Issues

Innovation sommeliers who are able to conceive a future based on innovation, must not only master one particular field but also have a bird's eye view of society by liberally moving across natural science, humanities, and social science. For this purpose, arts and science need to break free from a narrow, inward-looking outlook and create ways to complement each other.

After being invited by Doshisha University to set up their business school, I began teaching innovation theory and technology management at the university from the year 2004. Technology management is a business administration study that explores the methodology to create economic and social value by overcoming the barriers of humanities/social science and science, which mainly contains two aspects. One of these aspects is management study involving the analysis of elements and structure constituting this technology in order to find a new breakthrough method to explain the relationship with the market. The other aspect is management study of the elements and structure of the physics limits of technology to ensure that such technology never harms society. For both these aspects, if we do not delve into the source of innovation, we cannot find a solution.

I also started a business-related course where young minds could participate free of cost to take up the challenge of entrepreneurship. Participants are required to submit their business plans, and twenty of them would be selected to study entrepreneurship at a university abroad.

They would be sent to a different university each time, such as the University of California (UC) Los Angeles (UCLA), UC Irvine (UCI), University of Southern California, and UC Berkeley in the United States, EDHEC School of Business in France, and Cambridge University in England. These universities turned out talent, ranging from students who had soaked in the brisk entrepreneurial spirit of America, Britain, and France to many other successful entrepreneurs.

In the Japanese education system, science is considered as a technological tool and is only taught at the level to which it can benefit society. For this reason, natural science does not come under the purview of humanities and social science. Due to this, science literacy in Japan will never improve. I introduced a new subject to educate students on the role played by philosophy in the creation of science.

Science originated from philosophy and eventually branched out as an independent field to become one form that understood nature in its overwhelming diversity. Learning this process leads us to question Japan's past in importing science from abroad, and how these activities constitute transdisciplinary research, ranging from natural science to social science and the humanities.

However, at the university, an experience taught me how natural science has been isolated from the humanities and social science. This was at the business school that I had just set up. As the person in charge of designing technology management subjects as part of the educational curriculum, I made an announcement at the faculty meeting.

"Both in the semiconductor and nanotechnology industry, if we do not understand the technological aspects, we will be unable to plot out the industry structure or its vision. Therefore, quantum mechanics is important to understand this technology. It is for this reason I would like to teach the essence of quantum mechanics."

When I said this, an economics professor sitting next to me expressed his surprise and objected to my proposal.

"Quantum mechanics is a science that deals with topics like how momentum can no longer be determined once we determine the position and so on, which has absolutely no relation to the real world. Why would you think such a subject can add value to business studies? This is absolutely ridiculous." He began to get angry as he expressed his view. There were some professors who even laughed at my suggestion.

I was stunned as I did not quite expect this reaction. The microworld that deals with semiconductors, including mobile phones and computers, operates according to the principles of quantum mechanics. It is quantum mechanics that primarily supports the core of modern technology.

Economic experts do not understand this fact at all. This means that social science will not be able to draw a picture of the concept and vision of society in the upcoming future, let alone discuss the future of industrial society.

To clear the path to Japan's future, it is first important to again look at "what is innovation" to understand the first relationship between science and society.

When we work backwards through the chain of intellectual activities of innovation, we come to realize that industrial society has its origin in science, or in other words "knowledge creation." So, what sort of intellectual structure does the relationship between science and industrial society backed by science have? If we do not understand this fact properly, we will not be able to chart out the vision to create a new society.

After TEPCO happened to cause a nuclear disaster, the scientists involved began to express remorse at what had transpired without even making an attempt to reveal the truth to the citizens, and the citizens found themselves distrusting science even more. When actually faced with a situation where science was causing damage to society, the citizens witnessed how helpless the scientists were.

I believe that situation is the second relationship between science and society. These are none other than social problems or "trans-science" problems that science can cause but cannot be solved by science alone. I have addressed this argument from my own perspective in this book.

1.3   Structure of the Book

Let me briefly explain how this book is structured.

The second chapter compares Japan and the United States to get to the heart of the reality of the innovation crisis in Japan. The United States has moved out of the "large corporate central research laboratories" innovation model by introducing the "SBIR program" to support small businesses in the state and has been successful in creating a new innovation model. On the other hand, though Japan was seventeen years behind the United States in introducing the Japanese version of the SBIR program, it has been a spectacular failure. When we investigate the cause of this failure, we come across the differences in ideology within the program. Exploring what lies behind this ideology will help us illustrate what exactly is the innovation model of the 21st century.

In the third chapter, based on these considerations, I have examined what exactly is innovation, which I believe is the first relationship between science and society, and how science or knowledge creation is translated to the creation of economic and social value by undergoing some process or the other. I have introduced the "innovation diagram," which is an independent model outlining the principle underlying innovation, based on specific examples to shed light on the essence of innovation.

In the fourth chapter, I have discussed trans-science as the second relationship between science and society. As explained earlier, trans-science deals with issues that science can cause and which we can question science about, but science alone may not have the answer to this problem. Interestingly, this second relationship is compromised by the lack of innovation, which is the first relationship. After performing a causal analysis of the TEPCO nuclear accident of 2011 and the JR Fukuchiyama line accident of 2005, I have provided evidence to support this fact and explained why both of them are strongly linked.

In the fifth chapter, I have explored the background behind the demise of innovation while providing a general overview of post-war Japanese society. As a proposal on university reform to develop innovation sommeliers, I would like to throw some light on the course that Japan needs to take.

As mentioned above, the topics that have been covered in this book are wide-ranging, from very specific events such as the start-up support policies of Japan and the United States, and the nuclear accident to more abstract points of view like the structure for generating innovation and the relationship between science and society. The reader may be bewildered by the vast range of topics that I have covered. However, there is a common theme that largely connects all these aspects at the fundamental level: the simultaneous crisis in science and innovation.

Japan is currently continuing to move aimlessly without seeking the 21st century innovation model. Once we have the system in place, all we have to do is rescue these talented innovators who are adrift on a directionless boat. To do so, we have to fundamentally change the setup of the existing innovation system.

This book makes an attempt to provide a blueprint for the revival of Japan. Well then, let us set out on a journey to save the sinking ship.

Chapter 2

Why Has Japan Failed While America Succeeded?

2.1 What Are the Points of Difference Between Japan and America?

2.1.1 Decline of the Innovative Scientific Temper in Japan

The symbolic downfall of the Japanese electronics industry is not something we can merely attribute to corporate management strategy or to economic trends. In order to revive the science-based industries of Japan, we need to try and understand the structural factors behind this fall.

So, when and how did the international competitiveness of Japanese science-based industries begin to go downhill? And what kind of industrial fields are mainly witnessing this downslide?

In a resource-deficient Japan, there is no other way to increase national wealth than by importing raw materials, creating various kinds of value using science and technological capabilities, and exporting them at a price higher than that of the raw materials. That means that the value obtained by dividing the export price by the import price (this is referred to as terms of trade = export price/ import price) can turn out to be a good indicator to understand the dynamics of the international competitiveness of Japan. If the

Innovation Crisis: Successes, Pitfalls, and Solutions in Japan
Eiichi Yamaguchi
Copyright © 2019 Pan Stanford Publishing Pte. Ltd.
ISBN 978-981-4774-97-0 (Hardcover), 978-0-429-44862-1 (eBook)
www.panstanford.com

terms of trade is greater than 1, it means that the international competitiveness is high, and if the value is lower than 1, it means that this competitiveness is low. However, it is a relative value since it is measured based on a specific year.

It was when I spoke to economist Makoto Saito in 2014 (Kyoto Qualia Institute, 2014) that I learnt about this channel of thinking. Saito touches upon this method of discernment in his book on macroeconomics as well (Saito, 2014).

According to Saito, it was in the year 2000 that the trade terms began to drop rapidly, and the international competitiveness of Japan began to experience a sharp downturn. When the oil prices crashed after the Lehman shock, the trade terms improved a little, but in 2013, it became worse than the levels witnessed during the oil shock. This has led to a vicious cycle where the economy was forced to buy commodities at high prices and sell them at a much lower price, due to which Japan's income steadily made its way out of Japan. This enormous income seepage gathered speed after entering the 21st century, picked up more speed in the last few years, and now there seems to be no stopping it.

The export price is becoming cheaper than the import price because Japan's major export goods have lost their international competitiveness, and as export prices go down, imports are increasing at a fast pace as the foreign goods are no doubt superior.

With the exception of "mineral fuels" or petroleum, the industry that has been undermining the Japanese economy all along has been pharmaceuticals. The pharmaceutical industry has been in the red all along; moreover, as shown in Fig. 2.1, the amount of deficit saw a sudden surge in the year 2001 and exceeded 30 billion US dollars in 2014.

Looking at this trend, imports from overseas were almost constant at 6 billion US dollars until the year 2000 while the exports were also stable at about 500 million US dollars. However, in the year 2000, a certain "phase transition" occurred. From this point, imports from overseas exceeded 10 billion US dollars and rose suddenly to 30 billion US dollars in the year 2014 while the exports from Japan continued to remain at 1 billion US dollars.

The modern pharmaceutical industry ranks at the top of science-based industries where various disciplines intersect. This is because as the human genetic database is completed, a library of candidate compounds for drug discovery became available, and the methodology of drug discovery changed from "drugs to be discovered" to "drugs to be designed." According to this new methodology, it became possible to design high molecular weight proteins that were earlier not possible.

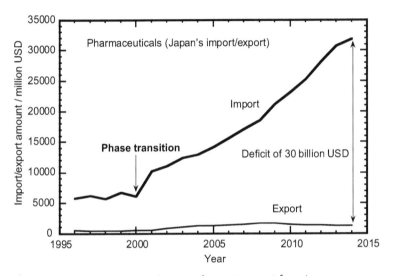

Figure 2.1 Import amount to Japan and export amount from Japan.

Science-based start-ups became all the more indispensable to optimize target molecules discovered through scientific research and creating the product precursor. Thus, the phase transition in methodology from "drugs to be discovered" to "drugs to be designed" has revolutionized the pharmaceutical industry as a business founded on science. Japan was completely lagging behind the world as far as this trend was concerned.

2.1.2 The Young Generation of Japan Robbed of Creative Opportunities

Let us approach the decline in scientific activity in Japan from another perspective (Iijima and Yamaguchi, 2015).

What I will focus on here is the fact that there has been no progress in the number of academic papers published in Japan ever since we entered the 21st century (Fig. 2.2). Looking at a breakdown by category in a hundred major academic fields, it is decreasing in

61 fields and rising in 39 fields. This is the "standstill state" based on the total normalized results.

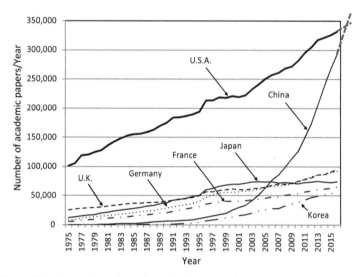

Figure 2.2 Annual number of academic papers from various countries as a function of year published. Here, the extrapolated lines for China and U.S.A. reveal China to be the world's best science powerhouse in 2018. Modified from Fig. 1 of Iijima and Yamaguchi (2015).

The problem is: In which fields are the number of academic papers declining? What stands out as we can see in Fig. 2.3 is that it has been falling steadily in the fields of physics, materials science, biochemistry, and molecular biology since the year 2004. If we look at physics in greater detail, although it is increasing in elementary particle physics, the decrease in applied physics and condensed matter physics is intense.

Applied physics, condensed matter physics, and materials science are, so to speak, fields of science that deal with substances and academic disciplines that support the semiconductor or nanotechnology industry and beyond that, the quantum mechanics industry. Biochemistry and molecular biology are academic fields, dealing in such subjects as iPS cells or immune regulation and genetic modification that are all indispensable to the pharmaceutical industry of the future, especially in advanced medical care.

On the other hand, in fields like chemistry, oncology, astronomy, mathematics, and the aforementioned elementary particle physics there has been a rise in the publication of academic papers. Except for chemistry, it cannot be said that these fields directly contribute to the major industries.

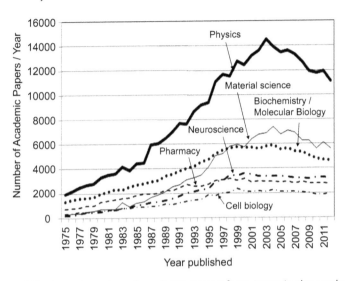

Figure 2.3 Annual number of academic papers from Japan in the academic field in which the number of academic papers from Japan tends to decrease, as a function of year published. Adapted from Fig. 2 of Iijima and Yamaguchi (2015).

In short, in Japan, science activities are rapidly declining in the most important fields directly connected with innovation that will support the 21st century.

How on earth did this happen? One reason is probably that the number of professional scientists working in private sector companies in these fields is on the decline.

Here, if we look at the trend in the number of graduate students in doctoral courses serving as a reserve force of professional scientists, as shown in Fig. 2.4, we can see that the figure for graduate students in the field of physics has begun to decline since 1997. Even for molecular biology we can assume that the results will probably be the same, so I will look at only physics here.

It takes roughly about six to eight years for graduate students to become professional scientists. If that is the case, it is consistent with the fact that the number of academic papers began to decline since the year 2004.

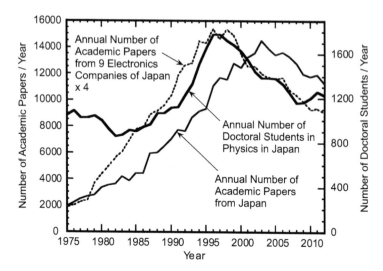

Figure 2.4 Annual number of academic papers from Japan (thin solid line), annual number of doctoral students in physics in Japan (thick solid line), and annual number of academic papers from 9 electronics companies of Japan (dotted line), as a function of year. Here 9 electronics companies are NTT, NEC, Hitachi, Toshiba, Mitsubishi Electric, Fujitsu, Panasonic, Sony, and Canon. Adapted from Fig. 5.1 of Iijima and Yamaguchi (2015).

So, why did the number of graduate students doing doctoral courses decrease specifically in the field of physics and applied physics, materials science, biochemistry, and molecular biology?

One conceivable reason is probably because the young generation has begun to think that even with a doctorate in these fields and with the opportunity of becoming an professional scientist, the future appears bleak. Even in the field of molecular biology, I have often heard that taking a doctorate degree leaves you with no job opportunities in Japan.

And it does not end here. The tendency of large corporates in Japan not to hire young scientists who have PhDs has become more and more evident since the late 1990s, and due to this, Japan has seen a substantial increase in the number of working poor PhD holders. Japan is about to end up as a society where creative young minds have no opportunities to create.

Among the Japanese science-based industries, the electronics industry, including the semiconductor industry, first began to downscale its scientific research activities significantly in the year 1996 (Fig. 2.5). Then, in around 1998, the pharmaceutical industry began to pull out of basic research (Fig. 2.6). The layoff of scientists from private corporations, their transfer from the research department, and the subsequent drop in the number of academic papers written by them corroborates this fact.

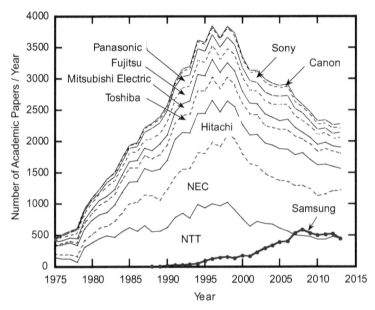

Figure 2.5 Annual number of academic papers from 9 electronics companies of Japan (NTT, NEC, Hitachi, Toshiba, Mitsubishi Electric, Fujitsu, Panasonic, Sony, and Canon) and from Samsung of Korea, as a function of year published. Adapted from Fig. 5.1 of Iijima and Yamaguchi (2015).

When we look at the number of physics papers of nine major electronics companies of Japan, we can see that it exactly meets the line depicting the downward trend in the number of doctoral students in the field of physics as seen in Fig. 2.4. If this happens, the "end of the era of central research laboratories" phenomenon will spread across industries, and it will end up also affecting the world of academics.

So, here is the thing. Since the early 1980s, Japan has been pursuing technological innovation with a focus on scientific research at the research labs of big corporations. However, large corporations were shutting down or downsizing their research labs one after the other in the latter half of the 1990s. Consequently, there was a significant reduction in opportunities for graduate students to showcase their creativity to society, leading the young graduate students to the conclusion that they would be unable to carry out research even if they joined private corporations. The number of young researchers who were pursuing condensed matter and applied physics, materials science, biochemistry, and molecular biology decreased at a steady pace, and as a result, the number of academic papers also dropped.

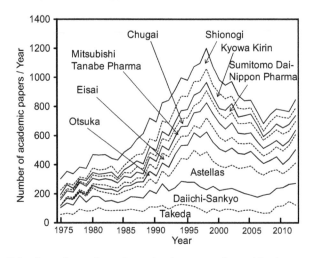

Figure 2.6 Annual number of academic papers from 10 pharmaceutical companies of Japan (Takeda, Daiichi-Sankyo, Astellas, Otsuka, Eisai, Mitsubishi Tanabe Pharma, Chugai, Sumitomo-Dainippon Pharma, Kyowa Kirin, and Shionogi). Adapted from Fig. 7.19 of Yamamoto and Yamaguchi (2015).

The withdrawal of Japanese private corporations that were responsible for eighty percent of research and development costs on scientific research deprived the creative young generation of their opportunities to create. Apart from causing the industrial competitiveness of Japan to fall sharply, it also reduced the competitiveness of science across Japan.

2.1.3 Are the Japanese Not Daring Enough?

This is where we are faced with a big question. The shutting down and downsizing of research laboratories is happening not just in

Japan but in America as well. Indeed the "end of the era of central research laboratories" was a development that initially took place in the United States.

In the early 20th century, the United States, which was a developing country for science, lured leading scientists from Europe to catch up with the situation. At the same time, they created a system where they could promote scientific research within private corporations, convert these results into technology for application in industry, and reinvest the generated profits back into research. Moreover, for the system to run smoothly, they brought together scientists and set up central research laboratories.

The first success story is the development of nylon by DuPont, and the other well-known "success" story is the Manhattan Project, which was an attempt to make the first atomic bomb.

The latter one specifically demonstrates the fact that if the world's best scientist group conducts basic research systematically, it can lead to the development of the ultimate weapon within a short span of time, and this was a key factor in determining the innovation policy of the United States after the Second World War. Finally, the transistor, which was one of the major innovations of the 20th century, came to be born in the AT&T Bell Laboratories..

However, after the recession between the 1970s and the 1980s, AT&T research laboratory first decided to pull out of scientific research in 1990 (Yamaguchi, 2006a). In 1991, the following year, IBM also began to pull out of scientific research. This move made by these two institutions that had driven the world of innovation caused a ripple effect all over the world. Following this, Xerox's PARC (Palo Alto Research Center) and Hewlett Packard also downscaled its basic research.

However, regardless of the fact that it was ahead of the world in downscaling basic research, unlike Japan, science- and technology-based industries did not see a downslide; rather they quickly started gathering momentum. This is apparent if we look at the growth in the IT industry and pharmaceutical industry in the United States after the 1990s.

Why did this happen? Solving this mystery appears to be the key to reviving Japanese science and industry.

Many of the scientists who moved out of corporate research institutions in the United States went on to either study further or establish their own start-ups. This had probably something to do with the "United States miracle" that brought about considerable economic revival in the 1990s.

In contrast, most of the Japanese researchers in corporates had to face the bitter prospect of being laid off or being transferred from the research department to production control or the sales department and so on, leaving them with no opportunities to utilize their talent.

A few of the researchers got out of Japan and took up jobs in Korean companies like Samsung, besides Taiwanese firms and even Chinese companies.

Though this movement received a negative backlash owing to the best brains and technology departing from Japan, from the perspective that the knowledge and skill of the innovators was being used to create value, it can be said that it has made a remarkable contribution to the world. Although it may have been a loss for Japan, Korea, Taiwan, China, and the rest of the world have benefited from this situation.

In fact, as shown in Fig. 2.5, Samsung that had neither the capability for scientific research or developing new technology until the 1980s significantly improved its organizational capability by taking in a whole lot of scientists and engineers from Japan in the 1990s and went on to become a global company with the world's best brains in the 2000s.

On the differences between America and Japan as mentioned above, there are many people who try to point out reasons such as the differences in national characteristics or cultural differences such as "unlike Americans who have set up their own companies, the Japanese lack entrepreneurial spirit" or "the Japanese do not have the courage to be independent, and are more large-enterprise oriented." One can also see people attributing the differences in culture such as American individualism and Japanese collectivism as another reason behind this trend. These opinions are what one may call the prevailing "stereotypical notions" on why Japan failed to innovate.

2.1.4 SBIR Has Dramatically Changed the Science-Based Industries of the United States

However, as someone who had been surveying America's Silicon Valley, the sacred ground for start-ups for more than ten years, I felt that the differences between Japan and the United States has to do with some or the other institutional factors rather than with the individual qualities of researchers or national character. I believe that these institutional factors go back to the 1980s, before Silicon Valley became a social phenomenon.

In fact, though I was a visiting scholar guiding physics graduate students at an American university from 1984 to 1985, I never came across any young students who said that they would like to establish their own start-up companies. Their dream was to join AT&T Bell Laboratories, and if that did not work out, stay back at the University and continue their studies as post doctoral fellows.

Therefore, this proves beyond any doubt that the United States acquired the new innovation model in the 1990s soon after it sought to move away from what is called the 20th century innovation model known as the "central research laboratories model of large corporations" in the 1980s. What is this model like?

When we go deeper into the details, what we come across is a program called SBIR (small business innovation research) started by the United States in 1982. An economics professor Akio Nishizawa drew attention to the SBIR program from early on saying that "the SBIR staff of the federal agency have been engaged in this business for a long time, so they have an excellent sense of judgment on technology, and we can say that they are very good at selecting start-up companies from the viewpoint of what kind of innovation is required for the nation" (Nishizawa, 2009). However, I had no idea about this. I got to know the impact of this system when I was investigating the reason why American start-ups that are obviously inferior in technology in next generation semiconductors manage to do way better than Japanese start-ups.

2.2 What Is SBIR?

2.2.1 The "Birth of a Star" System Set Up by the Government

So, what is the American SBIR program that laid the foundation for the new innovation model? To understand this, we need to go back to the United States of the 1970s that was drawn into recession as opposed to Japan that was making great strides forward.

Roland Tibbett (1924–2014), who was an executive in a large technology corporation and became a program manager for the National Science Foundation (NSF) in 1976 after setting up a venture capital firm, understood the reasons for the market failure plaguing science-based start-ups (Yoshihara-Yang, 2015).

In other words, since the risk of shaping scientific knowledge in a way that it becomes useful to society is very high, large corporations leave it to the market and refrain from investing. On the other hand, even if science-based start-ups feel inclined to providing their ideas to the world in a more concrete form, they cannot make sufficient investments as their net worth is way too small. Moreover, it is too much of a risk for private venture capital firms, so understandably they avoid investing in them.

In the development of advanced technology, the challenges that are difficult to overcome in the outcomes of basic research, implementation, and commercialization is referred to as the "valley of death." When such a finance gap is created, it will hinder innovation. However, as innovation ultimately swells the wealth and happiness of the citizens as a whole, this finance gap should be supplemented with public funds.

In this way, the seed of thought that "small businesses are the champions of innovation" sprouted in Tibbett, and he came to the conclusion that "the government must use national taxes, and initiate a program to correct this "market failure." So, he enthusiastically persuaded officials in the capital city of Washington D.C.

In this way, his idea came to bear fruit in the year 1982 with the passing of the Small Business Innovation Development Act, and the SBIR program was initiated in the same year under this act.

From the standpoint of a start-up company making an application, let me briefly explain the flow of SBIR.

In the United States, if you apply for SBIR and your application is accepted, you are given up to 150,000 US dollars as "award money" and an opportunity to build teams and create business models.

There are two kinds of "award money," namely "grant" and "contract." Grant, as the term indicates, is given as funds to support development and commercialization, and companies that are taken under the SBIR wing are allowed to take initiatives and use these funds at their disposal. On the other hand, a contract is a development contract like a subsidy in Japan, where the federal government takes the initiative and rigorously checks purchased items. While the projects from the Department of Defense (DoD) and the National Aeronautics and Space Administration (NASA) are mostly contracts, grants account for most of the dispersals from the Department of Health and Human Services (HHS), the Department of Energy (DoE), and the National Science Foundation (NSF).

If assessed as "feasible" at this first stage, you can get up to 1,500,000 US dollars as "award money" and take on the challenge of commercialization as the second stage. If you make further progress, the government may buy the future product that you have developed or introduce venture capital firms to you as the third stage.

More specifically, the American SBIR program is a "birth of a star" system that turns obscure scientists into entrepreneurs. Roughly put, it is a system in which the federal government contributes risk money from national taxes to turn a nobody or an obscure aspiring scientist, who has just started out and whose future potential is unknown, into promising scientists of the future.

2.2.2   The Three-Stage Selection Method

The SBIR program has three characteristics.

The first characteristic is that the law obliges the federal government to contribute a certain percentage of outsourced research funds to small business. Although this Small Business Innovation Development Act was an act valid only for a specified period of time, it has remained unchanged from the year 1982 to the present, and it has currently been extended until fiscal 2022 (October 2021 to September 2022) by parliament.

Due to this obligation, the eleven departments and agencies such as DoD, HHS, NASA, DoE, and NSF must lend a fixed percentage of the extramural research fund to SBIR. Most of the research expenses of HHS are being acted by National Institutes of Health (NIH).

This percentage was 2.5% between FY 1997 to 2011. After that, there was a 0.1% increase every year, and this percentage reached 3.0% in 2016. Moreover, after 2017, it was fixed at 3.2%.

The second characteristic, as mentioned earlier, is to determine the grantee of the award money using the three-stage selection method.

The first stage (Phase I) is where the feasibility of the idea is evaluated. The program managers/directors of DoD, HHS, or DoE present tasks such as "developing sensors that can help border security."

Graduate students who have started their own companies and young scientists such as postdoctoral fellows can apply. One in every six companies selected is given award money between 80,000 and 150,000 US dollars across a period of six months to one year. If selected, the feasibility is evaluated, and a simple knowledge of business administration is passed on.

The second stage (Phase II) is where an attempt is made to commercialize the technology. One in every two companies that have been highly rated in Phase I is selected and given award money of between 600,000 and 1.5 million US dollars across a period of approximately two years.

The average award money amount per company was 760,000, 730,000, and 700,000 US dollars in the years 2013, 2014, and 2015 respectively (Fujita, 2016). This amount, which translates to roughly 1 million US dollars, is just about enough to get past the "valley of death." If you get through Phase II, then you can proceed to the last phase, namely Phase III.

The third stage (Phase III) is in which the technology is actually commercialized, and the innovation is executed. In this phase, there is no award money, but the private venture capital firms are introduced.

In the case of DoD and DoE etc., the government departments and agencies procure the new product that is created. Since this is "something novel that is not currently available in the world" it is but natural that there will be no market as yet for the product. So, the government forcibly creates a market with the idea of boosting the growth of the companies selected under SBIR.

Thus: Competition (Phase I) to explore the possibility of business based on the idea of future technology → Visualization of competition (Phase II) to reduce the uncertainty of the risk money and convert the idea into a concrete business → Very specific business support (Phase III) through government procurement or introduction of venture capital firms.

This framework allows scientists, who have established their own companies by passing through this three-stage selection process, to change their mindset methodically from being researchers to entrepreneurs and take small steps toward becoming innovators.

2.2.3   Providing Identity as a Scientist

Well, the third characteristic is that the tasks presented by the program managers/directors are very specific. For example, "create a ceramic microprocessor capable of operating at ultra-high temperature," "create a device that can detect location information using wireless even in a forest zone," or "explore ion channel drug discovery using optical switches." In other words, it has a mission to "take on a challenge to make something exist that does not currently exist in the world."

The mission of program managers/directors is to prepare tasks that would facilitate the creation of such future industries and present these tasks to the applicants. To that end, program managers/directors are politically independent from researchers but must have the same in-depth and the most advanced knowledge of researchers.

Specifically speaking, the requirement is that they must have a doctoral degree, a year or more of research experience, academic papers written by them, and experience of having held positions above lecturer or associate professor.

It may be a bit late to bring this up, but let me define what exactly the term "scientist" means. Basically, scientist means a natural scientist or a person engaged in research on natural science. People who study humanities and social science are not called scientists, but they are usually grouped with people who are researching natural science and called "researchers." Even in this book, I have used the term "scientist" in this manner so far, and going forward, I shall use the term scientist when I refer to natural scientists.

Interestingly, in March 2014 in the United States, when I interviewed more than eleven program managers/directors who drew up the SBIR program plan and asked "What is your identity?" everyone with a PhD degree replied "I am a scientist." They said with confidence, "Since I am a scientist, I have a good sense of judgment."

In short, this is because in the United States, program managers/directors are also scientists. Since they are professionals who have acquired a PhD degree in natural science, have research experience, produce transdisciplinary research by bringing researchers together, and also design to create economic and social value from science, they are treated on equal terms with researchers and must have equivalent scientific knowledge.

At the same time, when I visited ten American start-ups that were selected under the SBIR program around the same time and asked their representatives "What is your identity?" almost all of them answer, "I am a scientist."

In other words, in the United States, entrepreneurs who have acquired a PhD degree in natural science and established start-ups through researchers also have the identity of a scientist.

This means that there are three types of scientists in America.

The first type includes researchers who carry out scientific research. The second type includes producers who explore scientific research and produce science-based industries. The third group includes innovators who actually create economic and social value from science. It is regrettable to say this, but the second category of scientists do not exist in Japan, and the third are not recognized as scientists.

From the start, the scientist profession has most often not been recognized as an occupation in Japan. For example, the Ministry of Education, Culture, Sports, Science and Technology's "Science and Technology White Paper" focused on "various professionals related to science and technology" for the first time since the 2003 edition. However, though there were "researchers" and "engineers" in the paper, there were no "professionals" called "scientists." I will touch upon this topic again in Chapter 5.

2.2.4 Building an Innovation Ecosystem

The eleven departments and agencies, such as the Department of Health and Human Services (HHS), invested a total of 38.9 billion US dollars in the SBIR program for a period of thirty-three years from 1983 to 2015. After 2008, over 2 billion US dollars have been spent annually for the SBIR program to the present day. The most recent budget for fiscal 2015 is 2.2 billion US dollars, and out of this, 43% is for the Department of Defense (DoD), 32% for the Department of Health and Human Services (HHS), 8% for the Department of Energy (DoE), 7% for National Aeronautics and Space Administration (NASA), and 7% for the National Science Foundation (NSF) (SBIR, 2016).

Since entering the 21st century, America has shaped more than two thousand obscure scientists into entrepreneurs of start-ups through this program every year. In this manner, 26,782 technological start-ups (263,530 companies including duplications) were born in the 33 years from the fiscal year 1983 to 2015 (Fujita, 2016).

In 2008, the National Research Council (NRC) conducted an assessment of the SBIR program and evaluated its results roughly as below (Wessner, 2008).

- ➢ The SBIR program has stimulated technological innovation. A variety of knowledge such as papers, patents, patent licenses, analysis models, and algorithms, was created.
- ➢ It helped in drawing research into the market and linked the university to the market. More than two-thirds of founders of companies selected under the SBIR program were university students, and about one-third of founders were university researchers before establishing their companies.
- ➢ It contributed to the commercialization of innovation. More than 20% of the technological start-ups were set up through the SBIR program. More than two-thirds of those companies selected under the SBIR program reported that they would have been unable to establish their companies had they not received SBIR funds.
- ➢ The federal government was able to make use of these start-ups to meet their research and development needs. In addition, the SBIR program fully met the procurement needs of various federal agencies.

> ➢ It has contributed in delivering innovation activities extensively to society. Between 1992 and 2005, every year, more than one-third of SBIR acquired companies were all new entrants.
> ➢ It encouraged participation of minorities and physically handicapped persons in technological innovation.

In this manner, in the United States, at the initiative of the government, a wide variety of related institutions in society such as universities and corporates acted autonomously to create an "innovation ecosystem" to accelerate the creation of innovation amidst an environment of competition and a complementary relationship.

2.3   Japan's Institutional Failure

2.3.1   Japanese SBIR Program That Ended Up as a Small and Medium Enterprise Support Policy

At this point, let us turn our attention back to Japan.

In Japan, even if you plan to set up your own company using the technology developed at the university, you have to use funds from your own savings for the initial capital investment. For this reason, it is almost impossible for young entrepreneurs with no savings to start companies of their own except for information technology that does not require investment funds.

If you seek financial aid from an investment company from the initial stages, you would lose your company ownership to the investment company, and the developers would most likely have to leave the technology behind in the company before they are eventually dismissed. Even in Japan, the implementation of a system akin to the American SBIR was long awaited.

Japan, which grew into a bubble economy from the latter half of the 1980s, fell into a major depression after the collapse of this bubble economy in the 1990s. Not only did the gross domestic product growth rate fall to minus 2% in the latter half of the 1990s, but the new business start-up rate in all industries eventually became less than the business shutdown rate.

In contrast, in America, the gross domestic product growth rate increased and crossed 4% between 1997 and 2000 and witnessed a boom. The reason for this is that start-up companies sprouted up one after the other, pushing up the economy as a whole, so the momentum to introduce a system in line with the SBIR program of the United States, which was viewed as a critical opportunity in this scenario, gathered strength in Japan as well.

Thus, the Small and Medium Enterprise Technological Innovation Program, which can also be called the "Japanese SBIR program" came into force from February 1999.

Initially, five ministries and agencies (Ministry of International Trade and Industry, Ministry of Posts and Telecommunications, Science and Technology Agency, Ministry of Health, Labor and Welfare, Ministry of Agriculture, Forestry and Fisheries) participated in the Japanese SBIR program. After that, the Ministry of Environment participated in 2001, the Ministry of Land, Infrastructure and Transport in 2005, and as of 2016, seven provinces participated, namely the Ministry of Economy, Trade and Industry, Ministry of Internal Affairs, Ministry of Education, Culture, Sports, Science and Technology, Ministry of Health, Labor and Welfare, Ministry of Agriculture, Forestry and Fisheries, Ministry of the Environment, and Ministry of Land, Infrastructure and Transport.

However, eventually, this Japanese SBIR program only looked similar to the American SBIR one but turned out to be quite different. The characteristics of the Japanese version are exactly the reverse of the American version.

The first characteristic differs from the American version in terms of the fact that "it is not obligatory to contribute a certain proportion of government outsourced research budget for small business."

Because it is not stipulated by law, the participation of the ministries is voluntary. Although this amount indicated a momentum toward closing in on the American version, with 1.19 billion US dollars issued in the fiscal year 2009, normally it stays around 200 to 400 million US dollars, which is one fifth to one tenth of the American version.

Moreover, this was nothing more than the existing subsidy system to which the name "Japanese SBIR" was merely affixed at a later point. Therefore, the subsidies are mostly "settlement payments," and since the inspection by the Board of Audit is also very stringent, it is a far cry from the "award money" of the United States.

The second characteristic is that it is not a multi-stage selection system with one exception (New Energy Venture Technology Innovation Program). Since the candidates for the subsidies will be asked about "achievements to date" to judge if they satisfy the criteria, young scientists such as graduate students and postdoctoral fellows with no proven track record, get excluded from the list.

Moreover, unlike the American SBIR, there is no creation of market for future products through government procurement and no introduction of venture capital firms as well. In this way, most of the target groups of companies selected under the Japanese SBIR program have turned into the existing small and medium enterprises.

The third characteristic is that they are not given any specific tasks to solve. In contrast to the American version, in the Japanese version only nondescript frameworks such as contributing to green innovation (strategy to transform into a low carbon society) are presented, and tasks with a more specific objective are not presented.

The administrators who assign the tasks do not have the capability to conceive topics that are relevant for the future in the first place. The reason is extremely simple. This is because there are no scientific program managers/directors in Japan that have the doctoral knowledge and experience comparable to researchers in the United States. The "good sense of judgment" to conceive the next-generation industry is absolutely lacking in people who are nurturing innovators.

As a result, unlike the American SBIR program, even if selected for the Japanese SBIR program, it is neither a matter of honor nor does it become a talking point because it is not accompanied by any significant benefit. For both selectors and those who get selected, it has ended up as an initiative with little meaning as is more of a chore that involves enormous administrative burdens.

2.3.2 Mimicking the United States without Understanding the Basic Concept

To start with, why did the "end of the era of central research laboratories of large corporations" happen in Japan after 1996? This is not due to the collapse of the land bubble in the early 1990s. The reason is that the research and development expenditure of each corporation rose rather than declined.

When we look at the trend in the number of papers by AT&T Bell Laboratories and IBM Research Laboratories in the United States, it can be seen that Bell Laboratories has reduced the number of papers since 1990 and so has IBM since 1991. The number of researchers has also declined greatly since the mid-1980s (Yamaguchi, 2006a). For them, as a corporate management policy, this meant a complete strategic change of stopping basic research and shifting to development.

Japanese corporations followed suit after accepting unquestioningly this trend in the United States and adopted the easy approach of "selecting and focusing" in order to withdraw from basic research in the name of management centered on "Shareholder value."

This phenomenon has also spread to the pharmaceutical industry in Japan (Yamamoto and Yamaguchi, 2015). If we look at the number of academic papers in Japan's pharmaceutical industry, as shown in Fig. 2.6, it has been decreasing after reaching its peak in 1998. Put plainly, this is nothing but withdrawal from scientific research. It can be said that this phenomenon followed the lead of electronics companies.

And what about the United States? If we look at the change over the years in the number of papers by pharmaceutical companies in the United States, it shows an upward trend as usual, as seen in Fig. 2.7 (Yamamoto and Yamaguchi, 2015). This means that the phenomenon "the end of the era of central research laboratories" that occurred in the United States is a phenomenon limited to the electronics industry, though in the pharmaceutical industry, it was a completely different strategy.

Japan imitated the United States and started the Japanese version of the SBIR program but did not finally understand the basic idea of the American SBIR program. As a result, the Japanese system, far from transforming scientists into entrepreneurs, simply ended up as a traditional policy with a "top-down" perspective in supporting small and medium enterprises.

So, what is the basic idea behind the American SBIR program?

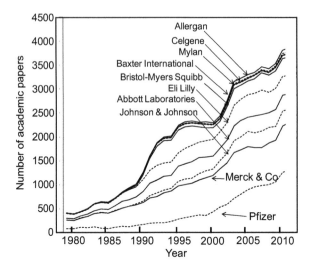

Figure 2.7 Annual number of academic papers from 10 pharmaceutical companies of U.S.A. (Pfizer, Merck, Johnson and Johnson, Abbott Laboratories, Eli Lilly, Bristol-Myers Squibb, Baxter International, Mylan, Celgene, and Allergan). Adapted from Fig. 7.18 of Yamamoto and Yamaguchi (2015).

2.3.3   Boldness of the American Industrial Policy

To establish science-based start-ups that can produce dramatic innovation, as Tibbett rightly understood, the higher the uncertainty, the greater the finance gap to bridge the "valley of death," and therefore, it is difficult to expect risk money to be supplied by the private sector.

As you can understand, the innovativeness of the American SBIR program lies in the idea of supplementing the finance gap that must be overcome in order to convert ideas that scientists have into innovation using public funds.

By this, scientists trying to be innovators can establish their own science-based companies without having to pay out of their own pockets from the conception stage itself. Moreover, as a by-product, scientists can create their own employment opportunities.

Research or the "knowledge creation" process is based on the inquisitiveness of scientists and does not essentially lead to the creation of economic value. The SBIR program has brought about a model that helps in channeling this process from the university research lab to the initiation of science businesses and assists scientists in creating specific value to society.

It opens the door to obscure, young scientists whose dream is to stay back at the academia as researchers to become innovators and take on the challenge of innovating something specific, without having to pay from their own pockets. It subtly tells us the fact that the self-realization of scientists can happen only if they convert their own "knowledge" to "value."

In other words, in the United States, "there is a goal in life apart from becoming a researcher" in aspiring scientists. For more than thirty years, it has continued using the policy of inviting scientists by using the motivation of encouraging innovation and offering the total "award money" of about 1 million US dollars.

Furthermore, the American SBIR program is a bold industrial policy launched by the nation based on the hypothesis that "small businesses give rise to innovation" and also a national project should be designed to nurture young obscure researchers into becoming entrepreneurs of start-ups.

Of course, the "small businesses" mentioned here simply does not refer to the existing small and medium enterprises. Scientists with the knowledge shaped at the academia level are selected as potential champions of new industry and honed to be entrepreneurs. In other words, rather than protecting existing small and medium enterprises, it is to create key players of the future industry from scratch.

Behind the idea that "small businesses give rise to innovation" is the perception that "large corporations can no longer innovate." The term "SBIR" itself reflects this perception. The people who created the system called SBIR affirmed this fact in 1982 when the U. S. economy was stagnating.

In that sense, it can be said that this policy itself is an innovation. It was based on the firm belief that "the future industry is born from new technology, that new technology comes from science, and that science is germinating among scientists."

In pursuing this, the federal government is nothing but an angel investor with a good sense of judgment. Moreover, the risk money that is contributed is not "investment" but an "award."

In this way, the SBIR became a new innovation model that could be aptly called "an organic open network model by science-based start-ups" that was created after the U. S. outgrew the "large enterprise research laboratory model."

This new organic open network model has a decisive point of difference from the erstwhile "large corporation's central research laboratories model." The difference is that while the latter is a self-sufficient system closed to only one company, in the former, networks that are made open to the world are formed autonomously. Networks that are open in this way have transformed not only the way society functions but also the way science functions.

2.3.4 74% of SBIR "Award Winners" in the United States Are PhD Holders

To understand the difference in the "basic idea" behind the SBIR program of the United States and Japan, let us look at where the representatives of the companies selected as part of the SBIR program come from. From what kind of academic discipline have they emerged and with what kind of knowledge have they brought technology to society? Specifically speaking, I checked their highest academic backgrounds and looked up the academic disciplines of these PhD holders.

In the American SBIR program, 63,648 scientists were selected in the thirty-three years from FY1983 to FY2015, and as mentioned above, 26,782 companies (263,530 companies including duplications) came into existence as technology start-ups. Since the beginning of the 21st century, on an average, it resulted in 2,390 companies coming up every year (Fujita, 2016).

Among them, I tracked down the career summary of each of the representatives selected in 2011 on the web. After studying them carefully for over a year, the number of representatives whose career summary I was able to trace on the web was 645.

If we break down the highest educational degree of these 645 people, we see that 73.7% are PhD degree holders, and apart from that, 12.1% of the total are undergraduate degree holders, and 14.3% are master degree holders.

As a method to systematically show the distribution of PhD degree holders, I will make use of the "academic landscape" (Fujita, Kawaguchi, and Yamaguchi, 2015) that we have developed.

In brief, this "academic landscape" shown in Fig. 2.8 is a visual representation of mathematically measured relationships of thirty-nine commonly known academic fields and is, so to speak, a map depicting the position of each academic discipline.

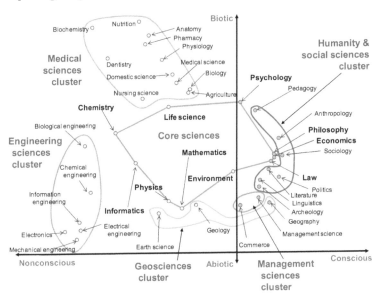

Figure 2.8 Academic landscape including 39 disciplines. Adapted from Fig. 6.3 of Fujita, Kawaguchi, and Yamaguchi (2015).

It has been created using the academic paper database "Google Scholar" to count the number of papers that simultaneously contain, say for example, mathematics and philosophy. Since this refers to the strength of the interaction between two academic fields, it can be changed to distance. If the number of papers that simultaneously contain mathematics and philosophy is equal to the sum of mathematics papers and philosophy papers, the distance between them is 0 (closest). Contrarily, if the number of papers that simultaneously contain mathematics and philosophy is 0, the distance between them is 1 (furthest). In this manner, the distance between the thirty-nine academic fields falls somewhere between 0 and 1.

This means that 39 points are floating inside the 39-dimensional space. Surprisingly, the 39 "flock of sardines" floating in 39-dimensional space are very flat, and if we take the appropriate two-dimensional axis for this, I realized that this distribution can be shown very well. In this way, it is the "academic landscape" of the 39 disciplines that is depicted in Fig. 2.8.

Interestingly, it means that the vertical axis is "biotic research"—"abiotic research," and the horizontal axis is "research focusing on the conscious mind" (humanities and social science)—"research that does not focus on the conscious mind" (natural science), and furthermore, all the academic fields in the first quadrant are humanities and social science while the academic fields in the second quadrant are natural science. This two-dimensional axis, when calculated using a computer, emphasizes that it is an outcome of such arbitrary distribution.

More interestingly, since mechanical engineering, electronics engineering, electrical engineering, computer science, biotechnology, etc. are strongly related to each other, it forms a closely grouped bunch (cluster), it is located on the side of "abiotic" and at the extremity of "non-conscious."

Ten academic disciplines (mathematics, physics, informatics, chemistry, life sciences, psychology, philosophy, economics, law, and environmental studies) are located near the center of the academic landscape. Five clusters (in clockwise direction, engineering cluster, medical cluster, humanities-social science cluster, business management cluster, and earth science cluster) are located around this area.

Let us name the ten academic fields placed in the center as "core science." These are academic fields that deal with the basic parts of each discipline, so to speak, and is also called "pure science." In Chapter 3, I would like to explain how important this core science is to effect a breakthrough.

Figure 2.9 shows the result of plotting the doctorate degree of applicants accepted under the SBIR program on this academic landscape (Yamaguchi, 2015a). The position of the circle is proportional to the academic discipline of the doctoral degree of the representatives, and the radius of the circle is proportional to the size of the share. From this figure, we can infer the following two things:

(1) Most of the fields the applicants selected under the American SBIR program come from core disciplines. The second is the engineering cluster, and the third the medical cluster. There are also some doctoral degree holders in humanities in fields such as psychology and philosophy.

(2) If we look at the academic disciplines of doctoral degree holders, the first position is chemistry (11.2%), and the second is physics (10.5%). Life sciences and biology collectively constitute 12.4%, and if we consider this as one field of study, it occupies the largest share (the denominator is the 645 applicants selected under the SBIR program whose career summary could be traced).

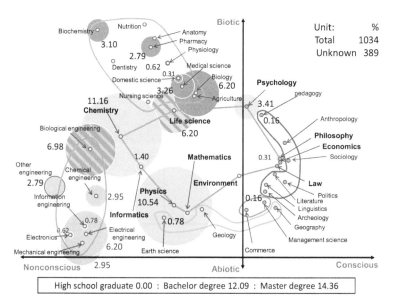

Figure 2.9 Discipline of PhD degree for the principal investigators of American SBIR (adopted in 2011). Adapted from Fig. 1.1 of Yamaguchi (2015).

From these two analysis results, we can see that the program managers/directors who run SBIR choose to assign topics that pure scientists can easily apply.

Even more, the United States systematically changed the advanced knowledge formed at university into innovation through the SBIR policy, and moreover, we can see that the U. S. government strategically considered core disciplines as most important for future innovation.

So, what kind of new industry was the United States thinking of creating through the SBIR policy?

If the 5,639 representatives are extracted in descending order of "award money" out of 46,354 people accepted under the SBIR program up to the fiscal year 2011, the coverage rate reaches about 40%. For each of these 5,639 representatives, we evaluated interaction with each of 39 academic disciplines, set the positioning in the academic field and plotted it on the academic landscape. The result is seen in Fig. 2.10 (Fujita, Kawaguchi, and Yamaguchi, 2015).

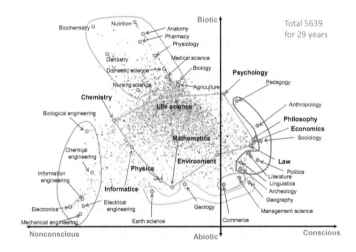

Figure 2.10 Academic discipline for the top 5,639 principal investigators of American SBIR for 29 years (from 1983 till 2011) (adopted in 2011). Adapted from Fig. 6.5 of Fujita, Kawaguchi, and Yamaguchi (2015) .

Then, the best part is that most of the points fell into the core academic discipline group. This shows that it is the scientists with pure science degrees who are establishing companies of their own. Besides, in pure science, the emphasis has been placed on life sciences and the second and third priority has been placed in either of the core academic discipline groups.

In short, it can be seen that the United States has tried to bring up the pharmaceutical industry strategically by creating the SBIR program based on the groundbreaking basic ideology that "innovation is born from science-based start-ups."

2.3.5   A Japan That Does Not Leverage University Knowledge

On the other hand, what is the highest academic degree of the representatives of the companies selected under SBIR in Japan? The number of companies selected was 23,339 in the twelve years between 1998 and 2010. Since this is an average of about 1,800 companies a year, it is comparable to the number of companies selected under SBIR Phase II of the United States.

The Small and Medium Enterprise Agency published only some of the company names. The names of the companies published were only those which had participated in the survey carried out by the Small and Medium Enterprise Agency, and this number stood at 3,559 companies, which was just about 15% of the total strength.

Checking the academic background of the representatives of these 3,559 companies to the extent possible using the database, I was able to find out the highest academic degree of 1,876 representatives. Moreover, for students who graduated from graduate school, I checked in the National Diet Library database whether they had obtained their PhD degree and their academic discipline, the results are shown in Fig. 2.11 (Yamaguchi, 2015a).

I would like you to compare this with Fig. 2.9. Contrary to the American SBIR program, only 7.7% of representatives obtained their doctorate since this program was implemented in Japan. Other than these, 20.6% of the total had passed junior high school, high school, technical college, and junior college. 68.2% were undergraduate degree holders, and 3.5% were master degree holders.

Moreover, if we look at the academic discipline of the PhD degree holders, PhD in engineering constitutes the largest percentage at about half of the entire percentage (3.4% out of the total 7.7%). The second is PhD in agriculture (1.4%), and the third PhD in medicine. On the other hand, the percentage of doctorate degree holders in the core academic group is only 0.6% in chemistry and only 0.5% in physics. There were no PhD degree holders in humanities/social sciences.

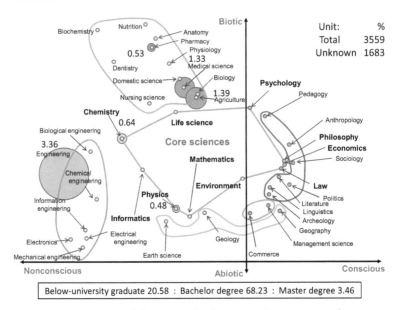

Below-university graduate 20.58 : Bachelor degree 68.23 : Master degree 3.46

Figure 2.11 Discipline of PhD degree for the principal investigators of Japanese SBIR (from 1998 till 2011). Here, I surveyed 3,559 companies published by the Small and Medium Enterprise Agency. Adapted from Fig. 1.2 of Yamaguchi (2015).

Comparing the proportion of doctoral degree holders, there is a big difference between the two countries with Japan at 7.7% as opposed to America at 73.7%. Moreover, while the United States has most graduates coming from the core disciplines or pure science, Japan has many graduates from applied science or practical science.

This also clearly shows that in Japan there is no strategic awareness on transforming advanced scientific knowledge born at university into innovation and creating a new industry out of it. Since "scientists" are not socially recognized, naturally there was no training process to transform them into innovators.

2.3.6 The U. S. Pharmaceutical Industry That Created High Additional Value

Let us examine what kind of variation was brought to society due to the differences in strategy and systems between the Japanese and American SBIR program.

First of all, how much did the American SBIR program, designed based on an extremely strategic perspective, actually lead to the success of science-based start-ups? In the above-mentioned "analysis of where the scientists come from," the focus was placed on biosciences (life sciences and biology), chemistry, and physics for companies selected under SBIR.

It can be said that the pharmaceutical industry, including the drug discovery start-ups created in the United States in the late 1970s, will be the source of innovation in the 21st century. If we look at the country-wise percentage in the global pharmaceutical market of 2011, the United States accounts for 36.2% of the world and is ranked No. 1 while Japan is at second place at 11.7%. By analyzing how much SBIR affected the pharmaceutical industry, we can measure the extent of the contribution of the SBIR program to society.

Firstly, when we look at "change in turnover of companies manufacturing insurance drugs" among the drug discovery companies originating in the United States, it can be seen that many of them have been selected under SBIR Phase II (hereinafter referred to as SBIR Phase II companies).

Of the total turnover made by the drug discovery start-ups in 2012 that accounts for 17% of the industry as a whole, the proportion of companies selected as part of SBIR Phase I and Phase II is close to 80% overall.

If we compare the number of companies and the total turnover, regardless of the fact that the number of non-SBIR companies is nearly three times that of SBIR companies, the total turnover of SBIR companies is more than three times that of non-SBIR companies. This shows how these companies have achieved satisfactory results in the pharmaceuticals industry.

In this industry, companies are not managed till the very end, and in most cases, they are acquired through M&A at some or the other stage. This accounts for the capital gain (Proceeds from M&A sales).

Over the last thirty years, SBIR Phase I and Phase II companies brought about 439.9 billion US dollars, combined with income gains (turnover) of 317 billion US dollars and capital gains of 122.9 billion US dollars in the pharmaceutical industry. This is, so to speak, the total value of the added value obtained through the SBIR program.

On the other hand, in the case of Japan, the turnover of companies selected under the SBIR program is negligible, and the total turnover stayed at just 110 million US dollars since the onset of operations in 1999. There are only four companies that have been selected under the SBIR program and have posted sales in the past. There was not even one company selected under the SBIR program that was acquired through M&A.

Needless to say, in the United States also, the SBIR capital comes from hard-earned taxpayer's money. Has this money been used effectively? To measure this, we need to look at the earnings from the SBIR industrial policy.

If we divide the total value-added amount brought about by SBIR companies in the pharmaceutical industry, (Turnover + Proceeds from M&A sales) by the total amount of SBIR award money contributed by the Department of Health and Human Services (HHS), we will be able to understand the SBIR gain. These results are shown in Fig. 2.12 (Yamamoto and Yamaguchi, 2015).

Totting up the total amount of SBIR award money contributed by HHS in the past thirty years, we find it was about to hit the 10 billion US dollars mark in 2013. As of 2013, thirty-one years after SBIR became operational, the revenue (Turnover + M&A amount) amounted to 45 times or more against government expenditure (Total SBIR budget spent by HHS). Due to the SBIR policy, the hard-earned taxpayer's money returned after multiplying itself 45 times or more.

For example, Amgen Inc. (Amgen), founded in 1980 by three chemists, has been incubating its gene cloning technology that was born shortly thereafter through SBIR, and the erythropoietin preparation (erythrocyte growth drug) developed from there became a blockbuster (a medicine that effects a breakthrough), bringing in about 18.7 billion US dollars in sales in 2013, and at 11th place in the global pharmaceutical industry.

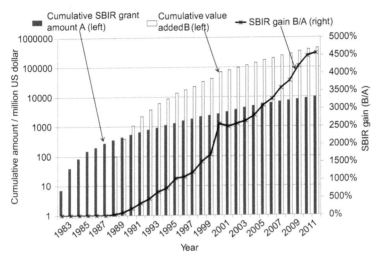

Figure 2.12 Cumulative amount of SBIR grant from U.S. Department of Health and Human Services (HHS), A, cumulative amount of value added (annual revenue + M&A sales value) from the companies adopted by SBIR, B, and the SBIR gain, B/A, as a function of year. Adapted from Fig. 7.11 of Yamamoto and Yamaguchi (2015).

Gilead Sciences, founded in 1987 by four young scientists, also incubated antiviral technology at SBIR, and the anti-influenza drugs and hepatitis C remedies introduced became blockbusters, bringing in a turnover of about 11.2 billion US dollars in 2013, ranking it at the 19th place in the world. The four young scientists were a medical doctor, Michael L. Riordan; a chemist, Peter B. Dervan, from California Institute of Technology; a biologist, Douglas A. Melton, from Harvard University; and a biochemist, Harold M. Weintraub, from the Fred Hutchinson Cancer Research Center.

2.3.7 The Japanese SBIR Program That Instead Lowered Sales

Finally, let us look at the "success rate" of companies selected under the American SBIR program. Here, the term "success" has been used if they have posted an annual turnover of one million US dollars at least once or have been successfully acquired through an M&A.

We find the success rate surprising with the success rate of companies in the SBIR Phase II growing year after year. It has consistently been over 1.2% since 2008 and as of 2012, it reached 1.29%. Although it may appear only 1% more, this success rate is high considering that it is for start-ups. For pharmaceutical research and development, this is a high percentage, just off by a digit.

Josh Lerner, an economics professor at Harvard University, has already made an econometric evaluation of the SBIR program in the United States (Lerner, 1999). In the ten years from 1985 to 1995, he measured how much the turnover of companies selected under the SBIR program increased on an average, and compared it with the growth in the turnover of companies that were not selected under SBIR.

In companies that were selected under the program, there was an average turnover increase of about 4 million US dollars per company. On the other hand, the average increase in the turnover of companies not selected under the program was about 1.1 million US dollars. In other words, this means that the companies that got selected under SBIR performed much better than those that did not by as much as 2.9 million US dollars.

So, how much has Japan's SBIR policy expanded Japan's wealth? To carry out the same analysis as Lerner, we studied the changes in turnover during the five years from 2006 to 2011 by comparing companies selected under the SBIR program and those that were not (Inoue and Yamaguchi, 2015).

As a result, in companies selected under the program, the turnover declined by about 2 million US dollars per company on an average, and in companies not selected under the program, the turnover dropped by about 0.7 million US dollars. This shows that the companies that were not selected under the program performed better than those that were by 1.3 million US dollars.

This is a very strange occurrence. Companies that received SBIR subsidies had a lower turnover than companies that did not receive these subsidies. So, what is actually happening?

This occurrence can probably be attributed to the fact that the national expenditure in the form of subsidies went into low performing small and medium enterprises rather than to the lack of effectiveness of SBIR subsidies. In any case, this shows that the Japanese SBIR program has not made good use of national taxes.

2.3.8 Starting Afresh with SBIR by Creating University Initiated Start-Ups

In order to encourage the setting up of university initiated start-ups in Japan, the knowledge cluster and industrial cluster policies have been implemented for a long time with an enormous budget. Unfortunately, most of these policies failed.

The reason is simple. This is because the subsidy was given to the university faculty and not to young scientists (graduate students or postdoctoral fellows) who established start-ups. These subsidies went up in smoke as research funds and did not create any new industries whatsoever.

As part of the "third arrow" of Abenomics, the government prepared a national budget of 180 billion yen in 2013 to promote the setting up of start-up companies in order to improve the Japanese new business start-up rate to twice as much of the current value.

Of these, 80 billion yen was invested by the Japan Science and Technology Agency and 100 billion yen by the University of Tokyo, Kyoto University, Osaka University, and Tohoku University. The government seems to have expected university professors and researchers to establish start-ups and transform science into innovation.

However, there is a fundamental flaw in the system design. Though a system for launching a venture capital firm in the university and enabling this company to make investments was put in place, the management team who assumed office in this company were amateurs who were not selected through open recruitment and had no prior experience of setting up their own technological start-ups or taking on the inherent risks and challenges.

This is totally different from the SBIR program director in the United States who examines the tasks (topics) for nurturing future industries across a year and designs the creation of new industries by embodying created "knowledge."

In the Japanese model, they do not have a "good sense of judgment" in innovation. There is no plan to implement a multi-level selection system like the American SBIR program. Moreover, it is not "award money" such as SBIR that can be freely used as risk money, but just "investment," and because sustainability is not guaranteed, it is completely different from the American SBIR program.

In general, it will take about five years or more for start-ups to show results. As shown in Fig. 2.12, in the U. S. pharmaceutical ventures also, it actually took seven years after the SBIR program was initiated that the SBIR gain (total revenue of the SBIR companies against the total SBIR budget spent by the Department of Health and Human Services) exceeded 1.

There will always be the moral hazard of receiving salaries from national taxes before parting ways. In a system where the country becomes an angel investor, people who execute this are required to have a strong will to do so without bias and favor.

Can we turn the 180 billion yen of national taxes into an opportunity to create a new industry in Japan without letting it go down the drain as before? Japan must hasten its pace to develop a renewed system design.

Chapter 3

How Is Innovation Born?

3.1 Abduction: Understanding the True Nature of Science

3.1.1 "Knowledge Creation" and "Knowledge Embodiment"

Beneath the surface of the SBIR program which helped the American science-based industries flourish is the ideology that "small businesses give rise to innovation."

To truly understand this ideology, we need to ask: "What is innovation in the first place?" In this chapter, I would like to examine the process to determine which scientific discovery creates economic and social value and get to the bottom of the principle underlying innovation.

In my book entitled "*Innovation: Paradigm Disruptions and Fields of Resonance*" (Yamaguchi, 2006a), I have shed light on the fact that innovation is created by a chain of activities involving "knowledge creation" and "knowledge embodiment."

"Knowledge creation" is firstly "knowing what nobody knows yet" and "seeing things that no one has seen before," both triggered by discovery. In other words, this is nothing but "science." For example, the discovery of the law of universal gravitation, the discovery of

Innovation Crisis: Successes, Pitfalls, and Solutions in Japan
Eiichi Yamaguchi
Copyright © 2019 Pan Stanford Publishing Pte. Ltd.
ISBN 978-981-4774-97-0 (Hardcover), 978-0-429-44862-1 (eBook)
www.panstanford.com

the relativity theory, and the discovery of quantum mechanics falls under this category.

Secondly, it is "making it exist which has never existed." This is in general, a part of "technology" and is referred to as "technological research," but if we use the words of scientist and philosopher Yoichiro Murakami (2010), it would probably be apt to call it "neo-type science." For example, the discovery of the transistor phenomenon, the discovery of the light emitting diode (LED) phenomenon, and the discovery of nuclear fission reaction (atomic bomb) are all "creations of knowledge." We generally refer to these intellectual activities as "research."

"Knowledge embodiment" refers to the intellectual activity of consolidating the scientific knowledge found through "knowledge creation" and turning it into something of economic and social value. This is called "technology" and if the discovery of the ferrite magnet is "knowledge creation," the development of magnetic tapes, floppy disks and hard disks is "knowledge embodiment." Therefore, "knowledge embodiment" may also be otherwise called "value creation." We generally refer to these intellectual activities as "development."

The relationship between the terms "research" and "development" and the terms "science" and "technology" that we often hear about is shown in Fig. 3.1. As you can see from this figure, "science" and "research" alone do not bring economic and social values, or, in general terms, "instrumental values." This means that "science is value-free and value-neutral," although science has "intrinsic values."

	Science	Technology
Development (Deduction)	×	○ To make values.
Research (Abduction)	○ To see what has never been seen.	○ To make it exist which has never existed.

Figure 3.1 Relationship between research/development and science/technology. ○ means "exist" and × means "not exist".

So, through what process is science that is "knowledge creation" converted into "knowledge embodiment" or in other words, economic and social "value creation?" Explaining this principle would mean not only ascertaining the essence of innovation that is created from the relationship between science and society but also involves delving into the fundamental truths about science.

At this point, I would also like to comment on "neo-type science." It was Murakami who pointed out that "science, which was once strictly limited to the scientific community but after entering the 20th century, has become science with moral responsibility to society because the interaction of science with society is triggered by the demands of society." The first is referred to as "prototype science" and the latter as "neo-type science" (Murakami, 2010).

However, in this phenomenological definition, the distinction between who the clients are and if they are related to society is unclear. In order to introduce clarity, in this book I have redefined "neo-type science" as "research" generally referred to as "technological research" because it involves "making it exist which has never existed."

3.1.2 "Day Science" and "Night Science"

Earlier in the book, I proposed the "innovation diagram" that is a unique model outlining human intellectual activities.

This diagram, as shown in Fig. 3.2, is a representation in two-dimensional space of "knowledge creation" and "knowledge embodiment" that are nothing but completely different intellectual activities on the orthogonal horizontal axis and vertical axis, respectively. I have used this two-dimensional innovation diagram to connect "knowledge creation" with "knowledge embodiment" and visually represent the process for producing innovation.

Going forward, I would like to use this innovation diagram to explain the essence of science and innovation and get to the SBIR program that I discussed in Chapter 2.

As shown in Fig. 3.2, in the two-dimensional innovation diagram, "knowledge creation" is depicted on the horizontal axis, "knowledge embodiment" on the vertical axis, and the area beneath the border drawn horizontally is considered as "soil." If we try to understand the dynamics of innovation using an analogy of a leaf bud sprouting from the soil, we will be able to understand the significance of this diagram.

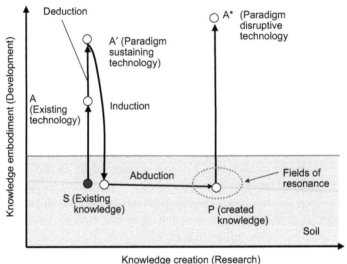

Figure 3.2 Innovation diagram. Adapted from Fig. 2.3 of Yamaguchi (2006) and Fig. 6.4 of Yamaguchi (2014).

The intellectual activity of human beings called "knowledge creation" occurs beneath the soil. The process of creating knowledge is like the path of an explorer who advances alone in true darkness without a candle. He/she has neither a map nor a textbook. He/she has to proceed further only by relying on the tacit knowledge within him/her.

On the other hand, above the soil is the world of sunshine. When the created knowledge begins to sprout from the soil, economic and social values are created. In other words, it will become a new product or service and bring some specific value to society.

Leo Esaki, Nobel laureate in physics, stated the principle of these two processes leading to innovation in another way (Esaki, 2007).

According to Esaki, science is "two-faced like the Roman God Janus" with the logos face and the pathos face. In order words, it has two sides, namely "day science" or "explicit knowledge" ("knowledge" that is verbalized) and written down in the form of textbooks etc. and "night science" or just "tacit knowledge" that is not yet verbalized.

It is by exploring new avenues alone without having any clues, and by repeated trial and error that one occasionally happens to find a brilliant breakthrough that is like a light at the end of a tunnel. It is said that all the sprouts of discovery that are propelling science forward are born out of such "night science." In other words, "knowledge creation" is "night science" and "knowledge embodiment" is "day science." although there are many scientists pursuing "day science" (Thomas Kuhn called this "puzzle-solving"), "day science" is not exactly science.

Innovation that leads to a breakthrough is always triggered by "night science" occurring beneath the soil. However, the people or society or the market that lives above the soil cannot see what is beneath it. Research conducted in corporates also lies beneath the soil and is invisible to the management. Its invisibility is eventually what leads corporates to stop or scale down their basic research activities.

Before getting into this, I would like to delve deeper into the facts related to our thought processes.

3.1.3   A Computer's Thought Process: Deduction and Induction

There are two main aspects to our thought process, namely "deduction" and "induction." "Deduction" is a method of reasoning where if hypothesis S is true, conclusions A and A′ can consequently be drawn. If the premise is "human beings are mortal" and if "Socrates is human," the conclusion that "Socrates is mortal" can consequently be drawn.

"Deduction" is a method of reasoning where a specific conclusion is reached based on some general premises, and we will invariably arrive at a "correct" conclusion. Although it is always true, it is established by a combination of existing knowledge, so it does not always create knowledge that no one had ever thought of before.

In principle, since computers can only deduce, it is natural that a computer defeats a human being in a game of chess, Shogi or Go. The modern-day computer is called a von Neumann type computer, and this is because the governing paradigm is that it functions based on some prebuilt programs that humans have created using deduction alone. The big data of Go is also nothing other than a mere combination of finite numbers 10^{400}.

Also, the software or system development is thus "knowledge embodiment" or "development" and is nothing more than an act of "deduction" to integrate the "knowledge" that needs to be consolidated. Therefore, in Fig. 3.2, it can be represented using an upward vector S → A or A → A′.

On the other hand, "induction" is the method of reasoning, where an attempt is made to infer general laws denoted as S from particular instances denoted as A. "Socrates, who is a human, is mortal". "Plato, who is a human, is mortal". Therefore, it is generalized as "All humans are mortal". In this case, incorrect conclusions can also be drawn if there are exceptions.

It is defined as the reverse process of "deduction," so in principle, this can be done by a von Neumann type of computer. However, to bring about the correct conclusion, we must refer to as many specific instances as possible.

Since we are able to store big data these days, it is not surprising that computers have become capable of communicating coherently with humans. Since "generalization of knowledge" is the reverse process of "knowledge embodiment," in Fig. 3.2, it can be represented using a downward vector A′ → S.

3.1.4 "Abduction" That Only Humans Can Do

The person who first presented the argument that the most essential intellectual activity in science is neither "deduction" nor "induction," but a third method of reasoning called "abduction" was American philosopher Charles Peirce (1839–1914) (Peirce, 1965). Peirce pointed out that:

> The surprising fact, S, is observed.
> But if a hypothesis P were true, then S would be a matter of course.
> Hence, there is reason to suspect that the hypothesis P is true.

Having said so, he referred to the method of inferring P from S as "abduction" and believed that "all scientific knowledge is a result of abduction." Since "abduction" is a method of reasoning that helps us make a hypothesis to best explain specific facts, it can also be considered as "hypothesis formation" or "hypothetical reasoning."

At first glance, we may not be able to distinguish "induction" from "abduction." This is because both are associated with deriving a general hypothesis from specific facts. However, the important point is that "abduction" brings about "knowledge creation," and as a result, the existing paradigm (natural view or world view) is defeated.

Astrologist Nicolaus Copernicus (1473–1543) believed that if we discard the extremely complex Geocentric theory and return to the heliocentric theory of planetary motion, not only will we be able to describe the planetary motion table more easily, but will also be able to logically explain the fact that the earth takes "one year" to revolve around the sun once. Thanks to "induction," we have been able to discard the geocentric theory and once again return to a generalization.

Encouraged by this inductive thinking, Johannes Kepler (1571–1630) discovered that the observations of planetary motion described by Tycho Brahe (1546–1601) throughout his life were a natural consequence if a certain hypothesis (Kepler's three laws) is assumed. The discovery of these three laws of Kepler was the first "abduction." In this hypothesis, S is the planetary observations of Brahe and the Heliocentric theory of Copernicus, and P is Kepler's three laws of planetary motion.

Not only this, Isaac Newton (1643–1727) discovered the astonishing fact that Kepler's three laws of planetary motion was a natural consequence if we assume the law of universal gravitation. The incredibleness of this Newton's "abduction" lies in the fact that not only the planets but the whole universe, including the distant stars. move according to the same law.

If we assume that only induction and deduction holds true for science, it would be totally unscientific to declare without proof that the law of universal gravitation extends to the entire universe. In Newton's hypothesis, S is Kepler's three laws of planetary motion, and although P is the law of universal gravitation, he did not know how hypothesis P was found. All we can say is that he probably had a flash of insight or a sudden revelation. Abductive reasoning is very much something that suddenly dawns on us.

When Indian mathematician Srinivasa Ramanujan (1887–1920) was asked by his patron Godfrey Hardy (1877–1947) at the

University of Cambridge "How do you get this intuition?" he replied "Goddess Namagiri puts formulas on my tongue while I sleep or pray."

In this manner, what can be found through "abduction" cannot be proved, just as it is impossible to deductively prove the Newton's law of universal gravitation. Although Ramanujan instinctively found numerous formulas and predictions that have been later proved by Hardy et al., "proof" is an act of deductive reasoning, and anyone with the required knowledge (therefore even a computer) can perform it.

However, no matter how much the computer evolves on the von Neumann paradigm, it will not be able to carry out the "abduction" that Newton managed to perform. This is because it is an act of creation that only humans are capable of, such as bringing to existence the things that do not exist in this world, or discovering things that nobody knows of.

Therefore, in principle, "abduction" is not programmable in the same way as it is impossible in principle to program "love and hate" or "conscience." The present-day computer, in any case is still imperfect and is nowhere close to the human mind.

Werner Heisenberg (1901–1976) describes the mental picture of the moment when "abduction" occurred, and he conceived quantum mechanics as follows (Hermann, 1976):

"In Helgoland, there was a moment in which it came to me as an enlightenment, when I saw that the energy was constant over time. It was pretty late at night. I figured it out and it was true. I climbed on a rock and saw the sunrise, and was happy (glücklich)."

3.1.5 Sustain the Paradigm or Disrupt It

Let us trace the process of producing innovation using the two-dimensional innovation diagram shown in Fig. 3.2 (Yamaguchi, 2006a, 2006b).

All innovations start from existing technology A based on a paradigm S (S → A). If you are responsible for the company's new product development, where would you try to proceed after starting from A? Naturally, it would be toward adding value by carrying out development while incorporating other technologies, or in other words, a vector in the upward direction (A → A').

However, in "deduction," where we embody knowledge with the aim of enhancing added value, we invariably reach a dead end when we go back and forth several times just in the way trees grow but eventually die and wither away.

For example, although Moore's law predicts that the number of transistors per square inch of integrated chips will become fourfold in three years (Moore, 1965), when the distance between electrodes becomes smaller than 10 nanometers, electrons become the waves of quantum mechanics, and we can no longer achieve greater speed even if the chips are made much smaller. It is predicted that we will run into a brick wall by 2020.

Also, silicon power transistors that appeared in Chapter 1: Introduction can withstand voltage only up to a certain limit. Besides, if we try to increase this forcibly, the electrical resistance becomes higher, and the power loss increases.

Also, no matter how much we try to increase the energy efficiency of the gasoline-powered engine, it cannot exceed the efficiency of an ideal engine called the Carnot cycle.

These dead ends are said to be the limits that accompany physics law (paradigm), which is otherwise called "physics limits." In the above three examples, each of the physics limits are in accordance with quantum mechanics, semiconductor physics, and statistical mechanics (second law of thermodynamics) respectively.

Let us call the innovation (A→A') that is produced in this way by repeated "deduction" according to a certain paradigm, which would eventually reach a dead end as a "paradigm sustaining innovation." This happens because it is based on the original paradigm and is nothing more than a continuation of the original paradigm.

Paradigm sustaining innovation is achieved by incorporating external technologies under a competitive environment and consolidating them to give shape to new technologies.

However, the sustainability of this competitiveness is not that high. This is because the new technology A' is based on the aggregation of "knowledge" that everyone can acquire in principle and can be easily replicated as long as there is no creation of new knowledge involved in the consolidation and integration process.

The patent system was created to prevent such imitations. The patent artificially protects the innovator, and the innovator tries to monopolize the market for a limited time period, thereby encouraging inventions. Although its significance in promoting development of the industry is high, since it is artificial, there are some organizations and countries that try to strategically circumvent it, so the cost that goes into preventing this from happening is none too low.

Besides, the paradigm sustaining innovation will invariably get stuck at some point due to the limitations of the paradigm that the initial "knowledge" originally contains. It not only is imitable but also has inevitable limitations.

3.1.6 Innovation Resulting in Breakthrough

However, even if the tree that is growing upward dies, if the soil is rich, then the growing roots would plough through the soil and put forth a new sprout above the soil. In the same way, if we get stuck while trying to create improved technology A′ from existing technology A, we will "induce" for the time being and get down to point S that is the "knowledge" that makes up A. (A′→ S)

So, we will drive "knowledge creation" with performing "abduction" in an unexpected direction. After rushing headlong into "night science," we discover a new paradigm P that no one had paid attention to. (S→ P)

After starting out on this new paradigm, by extending "knowledge embodiment" or the sprout of "deduction," it is possible to attain an entirely new value and new technology A* (P→A*).

Let us call such an innovation process (A→S→P→A*) of "induction" → "abduction" → "deduction" through the soil bringing about a breakthrough as "paradigm disruptive innovation".

The greatest paradigm disruptive innovation in the 20th century is no doubt the transistor. A transistor is an electronic device that operates on a paradigm called quantum mechanics as opposed to the vacuum tube functioning based on classical electromagnetism, and this encouraged the rapid development of electronics technology such as computers to start with.

A breakthrough happens initially only through paradigm disruptive innovation. In other words, the first action toward abduction would be to jump off from where we are to the soil of science.

Here, think about paradigm disruptive methods for the above-mentioned three paradigms. Firstly, in order to break through the physics limits of Moore's law in the integrated circuit, we need to step down to the "paradigm" that makes up the law, namely quantum mechanics, for now. This means examining the possibility of semiconductor devices that are based on completely different paradigms (e.g., MRAM). Currently, the world is trying to seek a solution with "more than Moore" technology.

Secondly, to break through the physics limits of the power transistor, we need to get down to semiconductor physics and look for materials that overcome the physics limits of silicon. This was GaN.

Speaking of the third example of the gasoline engine, it is the shuttle battery that I introduced in Chapter 1. It is a paradigm disruptive innovation breaking through these physics limits.

3.1.7 Process of Shuhari: Obeying, Detaching, and Leaving

If we consider the points so far, the process of paradigm disruptive innovation has the potential of being applied to all kinds of creative acts. For example, it is implicit in the ideology of "Shuhari: Obeying, Detaching, and Leaving" in Japanese martial arts or traditional arts.

"Obeying" is obedience or following the teachings of the teacher and learning that way "deductively" ($A \rightarrow A'$). "Detaching" is coming up with the antithesis of the teacher's teachings by "inducing" ($A' \rightarrow S$) and trying to break the old set of rules by "abduction" ($S \rightarrow P$). "Leaving" can be analogized as growing independent from the teacher and performing one's own "deduction" ($P \rightarrow A^*$).

Educational philosopher Tadashi Nishihira (Nishihira, 2009) argued that Zeami (1363–1443) described the process of obeying, detaching, and leaving in Noh as "Do imitate," "Do not imitate," "Can imitate." Nishihira gave me a very simple and easy to understand explanation (Kyoto Qualia Institute, 2015) during our conversation, so let me summarize it as follows:

Zeami described the dynamics of mastering Noh using three keywords, namely "Do imitate" $\rightarrow$ "Do not imitate" $\rightarrow$ "Can imitate." The first stage of being "Do imitate" involves mimicking the teachings of the teacher. The beginner tries to internalize these acts of imitation. The next stage is to "Do not imitate." This is a stage where the student no longer imitates his teacher but tries to "renounce conscious actions." Once this happens, the final stage of "Can imitate" occurs unexpectedly. In other words, a new language is created in a place that was once a "plane with no language."

Let us note that it is here that the innovation diagram contains an important hypothesis, that is, "abduction" as an act of "knowledge creation" does not occur unless "induction" happens first. "Abduction will not occur unless we go down to the essence of things in the process of "induction," and new knowledge that can disrupt the paradigm will not be created. "Abduction," in this way, happens only beneath the soil.

This means that the process of "detaching" that corresponds to "Do not imitate" actually consists of two processes, namely "induction" and "abduction." Pursuing "detachment" or to "Do not similar" would be to "learn the thesis properly, come up with the antithesis and then return to the point of origin".

To find out all the ways, there is no other course but to shift from "obeying" or "Do imitate." However, this alone will not be enough to understand the essence, so after coming up with the antithesis and returning to the point of origin, we would for the first time be able to "renounce conscious actions" or "Do not imitate."

3.2   Innovation Diagram of Blue LED

Now that we understand how to draw innovation diagrams in the two-dimensional space, let us practice it.

In the prologue, I made a case study explaining a remarkable innovation, the blue LED, which was created in Japan during the last moments of the twentieth century. Here, I will reconsider this innovation by creating an innovation diagram (Fig. 3.3).

3.2.1   Paradigm Sustaining Innovation by "Deduction"

(1) **Existing technology for LED → (upward arrow) ZnSe LED (paradigm sustaining innovation).**
There were chiefly two methods used in the creation of blue LED. See Fig. 3.4. This figure is plotted with various

semiconductors based on the value of the bandgap (vertical axis) and the distance of the atoms (horizontal axis). I will show the semiconductors that were able to emit light according to the bandgap using white circles and solid lines. In addition, I will show the semiconductors which were not able to emit light using black circles and dotted lines. Solid lines and broken lines indicate mixed crystals. For example, the solid line showing the connection of GaN and InN indicates $In_xGa_{1-x}N$ with x being 0 to 1.

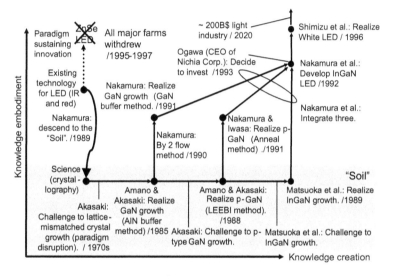

Figure 3.3   Innovation diagram for blue light-emitting diode.

There are also two semiconductors indicated by point P and point Q, which were the primary candidates for being able to emit blue light. Point P represents ZnSe and point Q represents indium gallium nitride ($In_xGa_{1-x}N$).

Which one seems probable from a technological standpoint? If you just check to see whether or not you can find a substrate material that can support the epitaxial growth, you will discover the answer that you are looking for. There is no lattice-matched substrate that has the same distance between its atoms for $In_xGa_{1-x}N$. Therefore, just by taking a look at Q, you will instantly know the epitaxial crystal growth will be impossible. On the other hand, ZnSe, shown through point

P, has the same distance between atoms as gallium arsenide (GaAs), and so it matches and meets the necessary conditions for the epitaxial crystal growth.

Now, which would you choose: GaN or ZnSe? Keep in mind most of the researchers who were taking on the challenge of developing the blue LED. Since they did not want to lose their footing with regard to the paradigm of the crystal growth, they must have chosen ZnSe.

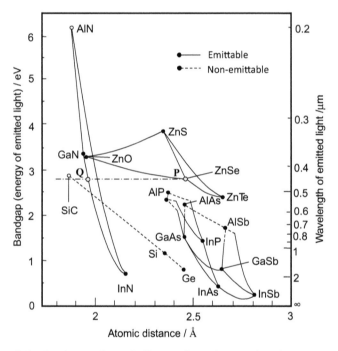

Figure 3.4 Bandgap and atomic distance for various semiconductors.

3.2.2 Challenge to Paradigm Disruptive Innovation by "Abduction"

(2) **Science (crystallography) → (right arrow) Akasaki: Challenge to lattice-mismatched crystal growth → Amano & Akasaki: Realize GaN growth (AlN buffer method)/ 1985.**

In 1973, Isamu Akasaki of the Matsushita Tokyo Research Institute decided to take on the challenge of growing crystals GaN. Knowing it was impossible to grow crystals using lattice-mismatched substrate with different distances between their atoms, he decided to go along challenging the paradigm of the crystal growth anyway. He chose sapphire as the substrate for the crystal growth of GaN.

However, what the Matsushita management did not take into consideration was Akasaki's drive to succeed. Akasaki decided to give up his current line of research. He then quit Matsushita and joined Nagoya University. As soon as he joined the university, he began conducting research on crystal growth of GaN.

A graduate student, Amano, expected that AlN might work as a buffer in between the substrate sapphire and GaN crystals. The GaN crystal could therefore be created. The discovery of the buffer technology by Akasaki and Amano created the first paradigm disruption.

(3) Realize GaN growth → Amano & Akasaki: Realize p-GaN (LEEBI method)/1988.

Akasaki and Amano also achieved the next paradigm disruption. At that time, no one had yet succeeded in creating p-type GaN. Then Akasaki and Amano added zinc (Zn) atoms to GaN and tried to transform, but not in vain. The Zn atoms that were added to GaN just would not act as acceptors (impurities which supplies electron holes).

In August 1987, Amano found interesting phenomenon when he observed Zn-doped GaN through an electron microscope. The light emissions seemed to be stronger after exposing electron beam to Zn-doped GaN. Amano expected that Zn atoms were activated by the electron beam.

Then, Akasaki and Amano, thought that magnesium (Mg) might be more effective in making p-type than Zn atoms. They tried exposing electron beam to Mg-doped GaN. They found that it turned into p-type. Thus, the second paradigm disruption, to create a p-type GaN, was realized.

(4) **Realize p-GaN → (right arrow) Matsuoka et al.: Realize In$_x$Ga$_{1-x}$N growth/1989.**

GaN emits ultraviolet rays. In order to emit blue light, one must first replace part of Ga atoms with In atoms.

However, no one knew how to add indium very effectively at that time. Everyone began to think that it was impossible. Matsuoka continued this experiment controlling the quantity of source gas and the growth temperature. However, it did not seem to work at all either. Then, Matsuoka looked at the basic thermodynamic theory for inspiration. He found out that it could work if the amount of ammonia as a source gas was increased by 16,000 times.

Then Matsuoka realized the success of the third paradigm disruption. Even now, the method, discovered by Matsuoka, has become the standard that is still used all over the world. However, NTT management ordered Matsuoka to stop his research in March 1992. After that, NTT did not look back after moving away from the paradigm disruptive innovation.

3.2.3 Accomplishment of Paradigm Disruptive Innovation by "Induction" and "Deduction"

(5) **Existing technology for LED → (down arrow) Nakamura: descend to the "Soil"/1989.**

Eventually, Nakamura, of Nichia, successfully created the blue LED, which would be used practically by combining these three paradigm disruptions. In 1988, he was allowed to conduct research in order to create blue LED because he had gained the trust of Nichia's founder, Nobuo Ogawa. Nakamura went to the United States and studied at University of Florida where one of his senior in university, Shiro Sakai, had been studying crystal growth technology, and then returned to Japan in April of 1989. Nakamura received three hundred million yen in a research expense budget from Ogawa to be used over four years, which allowed him to come up with the idea of the two-flow method in the summer of 1990.

(6) **Realize GaN, p-GaN, and In$_x$Ga$_{1-x}$N growth → (upward arrow) Nakamura et al. succeeded in catching up three paradigm disruptions → Develop InGaN LED/1992.**

The two-flow method, which is to input source gas into a reaction furnace from the side and at the same time inputting a great deal of nitrogen gas and hydrogen gas on the sapphire from the top, was introduced by Nakamura. He succeeded in growing a high quality GaN crystal on his first try of the two-flow method.

At first, he adopted the AlN buffer technology of Akasaki and Amano, and then he began to use GaN instead of AlN as the buffer. This was the second most valuable patent acquired by Nichia.

As no one else could, he wanted to create p-type using his own methods. Naruhito Iwasa, a new employee working under Nakamura, found that the resistance of GaN changed as he changed the quantity of hydrogen gas (which was used as a carrier gas).

Hydrogen atoms seemed to have a bad effect on the crystals. Iwasa put the crystals into nitrogen gas and simply annealed them. This worked beautifully in creating the p-type. It ended up being the strongest patent Nichia had ever acquired to date.

Nakamura thoroughly learned how to handle In from the methods Matsuoka had used. Matsuoka personally told Nakamura that "if you just follow my academic paper, you'll understand."

Thus, Nakamura completed work on the blue LED in September 1992.

(7) **InGaN LED → (upward arrow) Shimizu et al.: Realize white LED/1996.**

In 1996, Yoshinori Shimizu of Nichia, invented the technology which could emit white light by coating a fluorescent material on the surface of blue LED and, through trial and error, found out how to best create it. In that same year, the white LED was manufactured and then accepted to the color cellular phone market.

How did Nichia, which started out as a small venture business, become the biggest enterprise in the blue/white LED industry, even to the point of surpassing Toyota Gosei, which was based on the technology of Akasaki and Amano, and all the other leading enterprises in the field?

The reason is that the second president of Nichia decided to start manufacturing blue LEDs regardless of the fact that he was doing so against other management members' wishes in 1993. During that time, most enterprises did not recognize the potential that GaN held.

Enterprises have to take a huge risk to create a new market. As Toyota Gosei stood firmly in their position at the top of the industry, Ogawa's decision, which was really a do or die decision, literally saved Nichia from 1993 onwards.

3.2.4 Characteristics for the Paradigm Disruption of the Blue LED

There are common characteristics for paradigm disruptive innovation. Good enterprises, which have already a strong performance record, have to be careful of traps that might cause them to fall behind, or potentially fail. These traps come from the difficulty of transmitting tacit knowledge. In order to analyze it, I summarize the characteristics for paradigm disruptive innovation as follows:

1. There is no answer for paradigm disruptive innovation on extrapolating of existing technology A as "deduction". Large enterprises chose ZnSe as the material for blue LED, since it sustains the paradigm of the crystal growth , which is "lattice-matched condition." However, in the end, the key to realizing blue LED was not found on the crystal growth paradigm with ZnSe.

2. The new paradigm P (the buffer technology to disrupt the crystal growth paradigm) was discovered when one descends into the soil toward science S by "induction," and then continue "night science" by "abduction" beneath the soil. Only when one reaches the new paradigm P, one can accomplish

the paradigm disruptive technology A* (in this case, the blue LED).

3. Among the process, A → S → P → A*, the person of a good sense of judgment who is responsible for the success at the core, through sharing tacit knowledge, was none other than the founder of Nichia, Nobuo Ogawa. Finally, the person who finished the practically usable blue LED was not Akasaki, Amano, or Matsuoka, but Nakamura. Nakamura was able to conduct the research because of the Nichia founder Nobuo Ogawa's trust. Through this he ended up discovering the best conditions to grow GaN crystals, and developed blue LED by integrating the three paradigm disruptions provided by Akasaki, Amano, and Matsuoka.

3.3 Resonance and Transilience

3.3.1 Creating Fields of Resonance

Based on the above understanding of the innovation diagram, we can understand the facts related to the "end of the era of the central research laboratories" that occurred in Japan in the latter half of the 1990s.

In other words, the paradigm disruptive innovation of blue LED was a significant consequence of the central research laboratories in Japan, and "end of the era" meant the death of paradigm disruptive innovation.

Paradigm disruptive innovation follows the trajectory of creeping into the soil, groping in the dark, and sprouting from the soil.

Stopping research in fact means obstructing this abduction process by removing this soil. When this happens, we would no longer be able to go back to the initial knowledge using existing technology (creeping into the soil), and innovations that are produced would only be in the direction of improvising existing technology or paradigm sustaining innovation.

With scientists being laid off from organizations, there is nobody left who understands the process of "knowledge creation." Not only this, the know-how on the "abduction" process diminishes, vital clues on paradigm disruptive innovation are lost, and this is what led to the electronics industry and the pharmaceutical industry to completely lose their capability of producing breakthrough innovations.

This is also because both executives of organizations and bureaucrats in Japan were blind to the fact that it is knowledge creation through "abduction" that triggers paradigm disruptive innovation, and instead focused on the creation of direct and short-term value through "deduction."

However, the United States, in fact, strategically aimed for the next generation innovation model. This is exactly how one continues to carry out the paradigm disruptive innovation that I have been advocating.

So, what can we do to execute paradigm disruptive innovation?

The key to this lies in the presence of "fields of resonance."

A field of resonance is a real "place" where a person who considers "abduction" (knowledge creation) as his life goal and a person who considers "deduction" (knowledge embodiment) as his life goal, upon understanding and accepting the fact that these goals or existential drives are different, can strike a chord with other individuals and work with them. Each person's thoughts are different. Upon acknowledging this fact, you can share "tacit knowledge" by resonating with the sense of others and by sharing favorable and adverse experiences face to face.

In large organizations, the research department specializing in "knowledge creation" and the development department specializing in "knowledge embodiment" were separate divisions. Nevertheless, initially, spontaneous and autonomous fields of resonance were formed when these departments spent their spare time together.

However, when the objective of each department of the organization was clarified and "improved," the interaction between departments decreased, each of the departments became silos, and these fields of resonance collapsed.

The American SBIR program, as if it had an in-depth understanding of this situation, asked young researchers to "create fields of resonance" at the intersection of "knowledge creation" and "knowledge embodiment" and conveyed emphatically: "This is what the to-be society needs to look like." In other words, as introduced in Chapter 2, this is the implementation of the "birth of a star" system to transform obscure scientists engaged in "knowledge creation" to entrepreneurs who implement "knowledge embodiment."

In the SBIR program, scientists changed their mindsets, participated in the management of start-ups, and consequently created a social system that contributed to value creation.

3.3.2 The Pioneering Spirit of RIKEN Before World War II

At the same time, the US government recruited talented people who have the capability to conceive future paradigm disruptive innovations and apply them to social systems. These people are program managers/directors or "innovation sommeliers."

In the accomplishment of paradigm disruptive innovation, the process of "abduction" is imperative, and this is where scientists can contribute. The reason why the American SBIR program succeeded is because it encouraged scientists to not only become researchers but also transform themselves into entrepreneurs and program managers/directors and, consequently, they made the innovation diagram a social system.

When I visited the National Institutes of Health (NIH) in 2002 and asked the young postdocs at the institute, "What would you like to become in the future?" almost all of them answered "I would like to become a program director and not a researcher."

When I asked them the reason, they replied, "If I become a researcher, my research scope would be very limited, but as a program director, I would be able to discover technologies across disciplines, collaborate with researchers, and produce exciting innovations."

They are nothing but the scientists who, instead of functioning in isolation, have decided to conceptualize and chart out their own innovation diagram.

If we go back in time, it would be pertinent to recall the state of the Institute of Physical and Chemical Research (RIKEN) prior to World War II. RIKEN, set up in 1917, pioneered the 20th century type innovation model called the "central research laboratory model;" produced several eminent scientists such as Yoshio Nishina, Sin-Itiro Tomonaga, and Hideki Yukawa; and realized a number of scientific achievements.

RIKEN had two divisions, the Research Department pursuing pure science and the Development Department that created new products from the "knowledge" gained out of research, with the former allowing researchers the liberty to select their research topics. Researchers from different fields freely exchanged views, and there was an active interaction between different fields. Also, given its collaboration with universities, it served as a meeting place for young researchers. .

RIKEN offered an unrestrained research environment that Tomonaga called a "paradise for scientists" and built the fields of resonance required for the "knowledge creation."

Then, Masatoshi Okochi, the third-generation executive director who laid the foundation for RIKEN's golden age, encouraged basic science but at the same time did not neglect development to link the outcomes of research to industry.

The knowledge that was created at RIKEN led to the industrialization of several inventions including the formulation of vitamin A, "RIKEN vitamin" isolated and extracted from cod liver oil, ADSOL (adsorbent), metallic magnesium, synthetic sake, alumite, and many others, besides the establishment of several manufacturing companies equivalent to the start-ups of today.

3.3.3 "Transilience" or Knowledge Cross-Border

Before World War II, RIKEN not only promoted "knowledge creation" through "abduction" to produce paradigm disruptive innovation but also advocated the importance of the process of "transcending the borders of knowledge (knowledge cross-border)" to effortlessly bridge the barriers between academic disciplines through "transilience."

"Knowledge cross-border" that was briefly touched upon and in Chapter 1 and Chapter 2 is the intellectual activity of "discovering the essence of issues across disciplines, and finding a solution for these problems."

In the intellectual world, there are "borders" between disciplines, most typically the border between the humanities/social science and science. In organizations and in society, these become administrative and technological borders.

In addition, there are borders of academic disciplines. Among the sciences also, science and engineering have a very different interpretation of the world and different terminologies.

"Transcending these borders" presents people with the same difficulties as crossing the borders of the world. However, once we cross the border, there is a completely new interpretation of the world and basis of evaluation, and people gain freedom again.

Sociologist Anthony Richmond called such a person who can easily transcend borders as a transilient man (Richmond, 1969; Kawaguchi, 2004). Out of respect for Richmond, let me refer to this process of overcoming the limitations of academic disciplines as "transilience."

For example, American biologist James Watson (1928–present) and British physicist Francis Crick (1916–2004) discovered the DNA double helix in 1953 by transcending from biology to the world of physics. In the same way as the molecular system of heredity was discovered, a dramatic paradigm disruptive is sometimes created by knowledge that transcends academic disciplines.

Alternatively, scientists can gain completely new hints or ideas from the knowledge that philosophers have accumulated over time. In this manner, let us term this intellectual activity of actually "knowledge cross-border" from physics to biology and further to philosophy as "transilience."

For humans, this method of reasoning would mean to "leap into a completely different evaluation space." Even though it is a concept that represents the same phenomenon, different academic disciplines such as physics, chemistry, biology, economics, have different basis of evaluation and also use different terminologies. Since the groups of specialists constituting these fields are different, even the methodology they use to understand and objectively express natural and social phenomena varies considerably.

3.3.4 Leaping into a World with a Different Basis of Evaluation

If we represent the "transilience" activity within the innovation diagram, it can be extended further. We extend the two-dimensional diagram drawn using "knowledge creation" and "knowledge embodiment" and add a third dimension, namely "knowledge cross-border" as shown in Fig. 3.5.

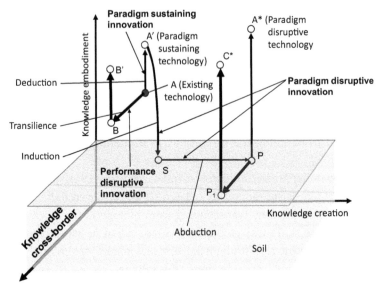

Figure 3.5 Innovation diagram extended to the three dimension. Adapted from Fig. 6.6 of Yamaguchi (2014).

In most cases, the intellectual activity of "knowledge embodiment" picks out "deduction" as the process in the consolidation of existing "knowledge" and adding value. This is the vector moving upward from A → A' in the figure. On the other hand, for "transilience," there is another vector in the direction of "knowledge cross-border," namely A → B. This represents the process of exploring a different basis of evaluation across different disciplines.

For example, in the development history of microprocessors, at a time when the pursuit of high-speed and high-performance was the mainstay of the industry, a start-up in Cambridge, U. K., set up by just eighteen people in 1990 deviated from the mainstream and ventured into the completely different direction of creating a low power consumption design and eventually went on to release a chip called ARM6 in 1990. On the other hand, Hitachi released a chip called SH1 in 1991 in pursuit of high-speed and high-performance.

Eventually, the ARM architecture (basic design) that lowers the performance and reduces power consumption went on to influence the mobile phone era of the future. This is a case where leaping into a world with a different basis of evaluation from the mainstream resulted in a totally new innovation. We will call this "performance disruptive innovation."

When "deduction," which adds value and improves technology, is dominant, it is just not possible for a leading legitimate enterprise to climb down a mountain that it has just scaled and jump into a world with a different basis of evaluation across different disciplines. Since ARM could not afford to take the high-speed, high-performance route, there was no other choice but to pursue a different basis of evaluation or the low power consumption route.

Now that the science has been subdivided into multiple disciplines, the barriers between academic disciplines obstruct transilience. In an age where it is impossible to survive unless innovation is accomplished, for Japanese companies that have lost the "abduction" capability over a span of 20 years to regain the "knowledge creation," it is critical for them to "knowledge cross-border" to overcome the barriers across academic disciplines. The key lies in nurturing people who can freely transcend social sciences and natural science.

3.3.5   iPS Cells Generated Through Transilience

"Abduction" is the essence of science, and "transilience" is what transcends the borders of knowledge. It is building up these two forces to point toward a direction that no one had ever considered before that would have been the model of science and innovation of the 21st century.

A good example of this model was seen in Japan in the year 2006. This was the discovery of iPS cells (induced pluripotent stem cells) by Shinya Yamanaka (1962–present). I would like to examine how this process extends across both "abduction" and "transilience" as seen in Fig. 3.6 (Yamaguchi, 2014).

We saw earlier that the DNA molecular structure with a double helix by Watson and Crick was discovered by transcending the disciplines from biology to physics (A → S). In 1958, Crick announced the hypothesis that "DNA is made up of a sequence with four kinds of bases (organic molecules), and this sequence makes up a digital code keeping life information to be carried to the next generation as genetic instructions." This "abduction" (S → P) is called central dogma, and it brings about paradigm disruptions to biology.

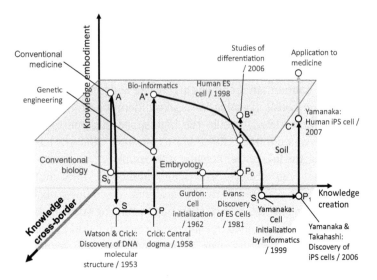

Figure 3.6 Innovation diagram for iPS cells. Adapted from Fig. 6.7 of Yamaguchi (2014).

As genetic engineering to read the nucleotide sequence of a gene was developed, a completely new drug discovery technology or a technology called bioinformatics was created "deductively" ($P \rightarrow A^*$) and the base sequence of the gene eventually came to be treated as information.

Until then, biology only pertained to the consolidation of natural knowledge based on experience accumulated over the years. From here, traditional medicine and pharmacology were "deductively" created through the integration of experience and knowledge ($S_0 \rightarrow A$).

Traditional biology will evolve on its own and will give rise to embryology that explores the mystery behind the origin of life. Following "abductive" findings such as cell initiation or embryonic stem cells (ES cells) that can differentiate into any tissue of the body ($S_0 \rightarrow P_0$), the mainstream of embryology eventually began to move in the "deductive" direction of the study of differentiation ($P_0 \rightarrow B^*$).

Now, Yamanaka dreamed of becoming a sports orthopedic surgeon and worked as an orthopedic researcher from 1987, but after two years he gave up and entered the Graduate School of Pharmacology to learn basic medicine. However, he got extremely frustrated with traditional pharmacology too and gave up this as well.

Nonetheless, Yamanaka, while pursuing his research on pharmacology, remembers being amazed by his encounter with a knockout mouse (a mouse in which a specific gene is inactivated to study the function of that gene) and realized that this could lead to a new breakthrough.

So, after obtaining his PhD in 1993, he went abroad to study molecular biology from scratch at the Gladstone Research Institute in the United States. Soon thereafter, when he cultured ES cells that killed the NATI cancer gene that he had discovered, he found that they lose their ability to differentiate into various types of cells. It was at that point that he became interested in ES cells that had been nothing more than a tool until that point.

After returning to Japan in 1996, he became a research assistant at Osaka City University Medical School and started studying ES cells from scratch. At the time, researchers from around the world were competing with each other to find out "what kind of cells were created from ES cells" in the area of cell differentiation, which was the mainstream of ES cells research as mentioned before.

Yamanaka, however, began a novel research, namely "creating ES-like cells not from live embryos cultured from fertilized eggs rather by using the genetic database." He did not know whether this was possible. However, if he was successful, he knew it would address the ethical problem of using fertilized eggs as well as the issue of immune rejection. If he failed, he would simply give up being a scientist and become a general practitioner. This was what he had decided when he took up the post of Associate Professor at the Nara Institute of Science and Technology in 1999.

In this way, taking the idea of Kazutoshi Takahashi (1977–present), in 2006, he picked four genes from the gene database and inserted them into the cells taken out with the virus and discovered that it creates cells that can differentiate into any tissue, namely iPS cells. Shortly after, he accepted a professorship at Kyoto University.

This achievement of Yamanaka, which won him the Nobel Prize in physiology or medicine along with the British biologist John Gurdon (1933–present) in 2012, shows that it is a major jump on the innovation diagram. This achievement, which disrupted the paradigm of embryology, has created a new field of study descending into the soil from a different academic discipline called bioinformatics, which is also completely different from the embryology of the past ($A^* \rightarrow S_1$).

Though Yamanaka suffered several setbacks, he moved across various fields of study like clinical orthopedics $\rightarrow$ pharmacology $\rightarrow$ molecular biology $\rightarrow$ cancer research $\rightarrow$ research on ES cells "amidst solitude." The discovery of iPS cells is "abduction" ($S_1 \rightarrow P_1$) as a result of "transilience."

Nonetheless, unable to settle down in one research field and moving quickly from one specialty to another, he remembers being uncertain about his future. At that point of time, he happened to attend a lecture by Susumu Tonegawa (1939–present) and shared this anxiety with him. Then, Tonegawa said as follows:

"Who said that continuity is important in research? As long as it interests you, do as you please."

In his words, I see the essence of "knowledge cross-border" through "transilience."

3.4 Breaking Away from Paradigm Sustaining Innovation

3.4.1 The Four Types of Innovation

Let me summarize the discussion on the source of innovation that I have explained so far.

The process of innovation can be classified into four types between 0 to 3.

"Type 0" is paradigm sustaining innovation and is shown as a process of A $\rightarrow$ A' in Fig. 3.5 or Fig. 3.7. It is a process that is established only through deductive reasoning, and it is an innovation accomplished by improving existing technology without changing the paradigm and integrating or incorporating other knowledge. Providing a camera function to a mobile phone etc. falls under this type. Since "Type 0" can be achieved by integrating existing knowledge, it is possible to draw a road map.

"Type 1" is paradigm disruptive innovation. This type of innovation entails looking at the future of existing technology A and judging if we need to aim for goal A* in extension A' or whether it is a dead end, and if we find out that it is a dead end, it is the process of creeping into the "soil" and choosing A′ → S → P → A* shown in Fig. 3.7. The invention of the transistor and blue LED fall under this type (Yamaguchi, 2006a).

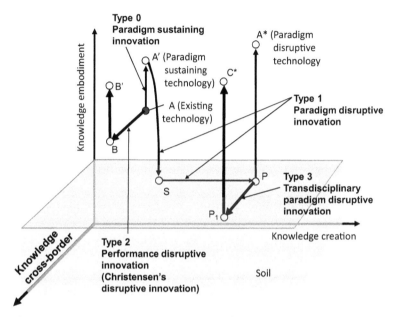

Figure 3.7 Innovation type 0 to type 3 are shown in the innovation diagram extended to the three dimensions. Adapted from Fig. 6.6 of Yamaguchi (2014).

"Type 2" is performance disruptive innovation. "Knowledge embodiment" does not mean carrying out development in the direction based on the existing basis of evaluation but asking the question as to what can be the basis of evaluation, discovering the new basis of evaluation A → B that the future society is seeking, and choosing the process of A → B → B′ shown in Fig. 3.7. The development carried out by ARM focusing on the basis of evaluation of low power consumption and not high-speed and high-performance in the development of microprocessors falls under this type of innovation.

"Type 3" is transilience or transcending the borders of academic disciplines beneath the soil in Fig. 3.7 (P→P₁). Every academic discipline has an inevitably unique basis of evaluation, and not only this, each of the groups involved has a unique language and culture, which are barriers preventing transilience.

Even in such a situation, by transcending the barriers across academic disciplines, it is innovation obtained through the process of $P \rightarrow P_1 \rightarrow C^*$. This innovation can be called transdisciplinary paradigm disruptive innovation. Discovery of iPS cells by Shinya Yamanaka and the innovation of entirely new regenerative medicine that he created based on these findings is a typical example of Type 3 innovation.

However, innovation that was called as "disruptive innovation" by the world's foremost authority on innovation research, Clayton Christensen, falls under Type 2 (Yamaguchi, 2014).

Christensen quantitatively examined the history of hard disk drives and called the innovation of increasing the recording density per square inch or "increasing performance" as "paradigm sustaining innovation."

On the other hand, Christensen called the act of reducing the storage capacity by dropping the size 14 inches → 8 inches → 5.25 inches → 3.5 inches by the start-up company that eventually led to the downfall of large corporations as innovation for "reducing performance" or "disruptive innovation" (Christensen, 1997). It is being perceptive of the fact that owing to management decisions, large corporations would get caught in an unavoidable value network and go down.

Subsequently, I demonstrated that his argument to classify a transistor as disruptive innovation is incorrect and said that innovation that Christensen referred to as disruptive innovation should be called "performance disruptive innovation." (Yamaguchi, 2003; 2006). This is because the disruptive innovation of Christensen is nothing but changing the basis of evaluation for performance, discovering a new basis of evaluation that the society of the future seeks, and choosing a process toward a "different future."

3.4.2 Innovation Sommeliers Are Required

The hypothesis of a "mountain climbing trap" applies to Christensen's disruptive innovation or Type 2 "performance disruptive innovation." Let us take the example of the competition between Hitachi and ARM as a typical example of Type 2.

This was the competition between Hitachi's high-speed and high-performance chip and ARM's chip with low-speed and low-performance, but low power consumption as well. Although from the standpoint of computer science, the design ideology of the latter meant completely going astray, it was this strong sense of vision that "We would carry computers in our palms" that steered them off this regular course.

Eventually, the era of mobile phones accompanied with the "different future" presented itself, and the ARM chip became the new norm in the microprocessor of mobile phones.

Thus, although Hitachi was climbing up toward the high-speed, high-performance market basis of evaluation that was mainstream at that time, it was unable to get down to a "different future" of low power consumption that was the new basis of evaluation. Hitachi also developed a low power consumption chip called SH3 four years after ARM's chip and released it in 1995, but it was already too late. Also, Hitachi's policy of "in-house production" backfired, and it completely lost out to ARM's open innovation strategy of licensing the microprocessor circuit diagram to everyone.

It is no exaggeration to say that innovations of the majority of large corporations fall under the category of Type 0 paradigm sustaining innovation strategy. Since soil is not needed, management finds it easy to take decisions and exercise control. However, one must not forget that Type 0 includes two major pitfalls.

The first pitfall is when Type 1 "paradigm disruptive innovation" makes its appearance in the world. The other is when Type 2 "performance disruptive innovation" (Christensen's disruptive innovation) makes its appearance. Regardless of both the events, history tells us that if we handle it after it shows up, we will by never be able to go down from the Type 0 mountain and will have no choice but to quit the market.

So, what should companies do? In order to deal with the former, we need to keep constant watch beneath the soil and have a team performing "technology intelligence" and exploring where paradigm disruption is occurring in the world. This is the "innovation sommeliers" team having a good sense of judgment. The era in which corporates can function without this team is long gone.

In order to deal with the latter, it is necessary to keep a team that always conceptualizes the vision of a future society. The "innovation sommeliers" team should be responsible for this as well. Therefore, first and foremost, this sinking Japan needs to nurture people who can freely transcend not only natural sciences but also social science, and humanities.

Let us look at this again in Chapter 5.

Chapter 4

Science Resonating with Society

4.1 What Is Trans-science?

4.1.1 "Trans-science" Problems That Science Alone Cannot Solve

In the previous chapter, to understand the essence of innovation we used the innovation diagram to discuss the process through which science or "knowledge creation" is translated into the economic and social "value creation."

Innovation is a creative act that is produced by the collaboration of science and society. If we consider this as the first relationship between science and society, we must not forget that the second relationship also exists between the two. In short, these are "trans-science" problems that science can cause but cannot be solved without the involvement of society. Many of these problems are instances in which science inflicts damage on society.

Trans-science is a concept advocated by American physicist Alvin Weinberg (1915–2006) in 1972 and is defined as "problems for which questions can be asked of science, and yet which cannot be answered by science" (Weinberg, 1972). Since this questions "transcend" science, he used the prefix "trans," and referred to it as "trans-science."

Innovation Crisis: Successes, Pitfalls, and Solutions in Japan
Eiichi Yamaguchi
Copyright © 2019 Pan Stanford Publishing Pte. Ltd.
ISBN 978-981-4774-97-0 (Hardcover), 978-0-429-44862-1 (eBook)
www.panstanford.com

As mentioned in the previous chapter, science is value-free and value-neutral, and the truth discovered in this process is the creation of new knowledge. Therefore, science by itself does not bring about economic and social values, or, in general terms, "instrumental values."

Society has not only used the knowledge created through science to produce various kinds of technology but has also created the values to make decisions about politics and policy. However, a field that is inextricably linked to both science and politics (decision making in society) has come to exist, that is, the field of trans-science.

For example, TEPCO caused a serious Level 7 nuclear accident at the Fukushima Daiichi Nuclear Power Station due to the tsunami of the Great East Japan earthquake on March 11, 2011. The opacity of the Japan's "nuclear village" comprising the industry, government, academia, and media involved in the nuclear power industry became a topic of considerable discussion, and this triggered a debate within Japan about the need to review the relationship between science and society.

Until this point, since there was no "platform" for discussions on trans-science, scientists and citizens did not open up about these issues. Since science did not interact with society, citizens thought that it was better to leave it to the experts of this field, that is, the scientists.

However, going forward, science should exist for the sake of society. For instance, life scientists should not only focus on "discovery" but also on the "health of human beings." However, the current scenario where science is causing severe damage to society, as in the case of the nuclear accident of TEPCO, has stirred debate about the need to solve these problems through democracy, based on which modern society takes decisions.

These are indeed "problems for which questions can be asked of science, and yet which cannot be answered or solved by science alone." In other words, trans-science is a concept of paradox, which includes the democratic society that moves ahead based on majority rule and science in which the majority rule does not hold any significance.

This second relationship between science and society is impacted by the lack of innovation, which is the first relationship. In other words, the deterioration in innovation capabilities of corporate organizations leads to not only reduced profit or decreased competitiveness, but also undermines the ability of organizations to manage a crisis, which in turn can also cause irreparable damage to society.

To empirically discuss this fact, we will analyze the new relationship between science and society in this chapter and also review two major accidents that occurred in the recent past. Before we discuss the details, let us take a closer look at the past discussions on trans-science.

4.1.2 Republic of Trans-science

As Weinberg pointed out, "problems for which questions can be asked of science, and yet which cannot be answered by science" can be put into three categories.

The first category is problems for which science cannot give any feasible answer, such as "impact of low level of radiation on living organisms" or "probability of occurrence of extremely low probability events" and "engineering." For example, it is difficult to estimate the probability of multiple emergency devices of a nuclear reactor failing to function at the same time, and when we introduce a new technology into society by engineering, we need to piece together our incomplete and fragmentary scientific knowledge and take decisions.

The second category is unpredictable problems that lack logic, such as "human and social behavior." Therefore, social science is considered trans-science.

The third is associated with the evaluation of the "value of science." For example, when it comes to deciding "whether to give priority to pure science or applied science," we must address "value" rather than the "truth" of science, and therefore, we need to make judgments by transcending science.

Weinberg explained this fact using the phrase "republic of trans-science" which is a term modeled after the scientist philosopher Michael Polanyi's (1891–1976) concept "republic of science" (Polanyi, 1962). His argument is as follows:

The validity of the knowledge created by science is established and maintained through the critical judgment of scientific peers. On the other hand, there are no opportunities for citizens to be involved in the evaluation of science created by scientists. The citizens of this "republic of science" are scientists, and only scientists are allowed to express themselves here.

However, the "republic of trans-science" consists of two kinds of members: (1) scientists; citizens of the "republic of science" and the (2) citizens of the actual society; the people who discuss the politics of this republic and make decisions for the country.

However, in a non-democratic society, such decisions are not very accessible to the citizens. Therefore, Weinberg provides the following example to demonstrate that the nature of the latter is extremely important to the discussion of trans-science problems.

In a democratic society like the United States, the possibility of nuclear accidents is frequently discussed with citizens, and consequently, safety and emergency equipment are installed so often that it even puzzles the experts. However, in a non-democratic society such as the Soviet Union, since citizens do not have the right to participate or lack the information on such scientific and technical matters, it is said that the containment vessel was not installed in the pressurized water reactor.

Weinberg presents the following conclusion based on the above discussion.

The responsibility and role of scientists are to clearly explain the end of science and the beginning of trans-science. Since scientists have knowledge of this border, they must utilize all their scientific knowledge to control the chaotic discussions on trans-science. Scientists must also promote the participation of citizens in such discussions.

However, there is one problem in this intense and elaborate discussion, that is, the line between science and trans-science is not strictly defined.

Although it is the responsibility of scientists to determine where to draw this line, they hesitate to interfere in these issues. On the other hand, it leads to a situation where society does not encourage participation of scientists even in the case of trans-science problems.

4.1.3   Civilian Control of Science and Technology

Biologist Atsuhiro Shibatani (1920–2011) was the first to react to Weinberg's argument in Japan. In his book *Han kagaku-ron* (Anti-science Theory) published in 1973, he presented Weinberg's argument and added the following points (Shibatani, 1973).

> In Japan, since politicians can step into the inner world of science relatively easily without much opposition, there are limited opportunities to have sufficient discussions within the scientific world. Consequently, the possibility of giving a straightforward definition for this border between science and trans-science is also extremely limited, and you could say that there are hardly any opportunities for scientists to make a contribution from a neutral standpoint. Instead, even if there are few scientists who make such an attempt, the rest of the scientific world would mostly avoid or refrain from carrying out scientific discussions, or try and suppress this debate due to specific political clout or to show public or implicit solidarity.

If we read Shibatani's work today, we have to say that he had correctly foreseen the gravity of the "nuclear village" which was exposed by the TEPCO Fukushima Daiichi nuclear accident, almost forty years ago.

Subsequently, in his book "*Toransu-Saiensu no jidai* (The era of Trans-Science)" (Kobayashi, 2007), scientific philosopher Tadashi Kobayashi re-evaluated Weinberg's argument and discussed the current importance of trans-science from the perspective of science communication, which is his specialty.

Kobayashi criticized conventional science communication as being a "deficit model." In short, it means "citizens lack knowledge of science, and the differences of opinion with scientists is due to the lack thereof." Therefore, science communication thus far was considered as a means of providing scientific knowledge to ignorant citizens.

However, in the present day and age where trans-sciene is continuing to grow, he stressed on the need of a "dialogue model" or the form of communication in which citizens and experts learn from each other through interactions instead of a one-way communication. To achieve this goal, he concluded that it is the "consensus conference" between citizens and experts based on the "dialogue model" that would prove to be an extremely effective method.

However, Kobayashi stipulated three conditions to conduct this "consensus conference."

(1) Do not aim at consensus building. This is because bringing aspects that experts themselves cannot agree on to citizens will only cause confusion.

(2) The discussion must be "rational."

(3) Experts from social science or humanities must participate in the discussion.

Under these circumstances, Kobayashi asserted that "further consideration can be given to the concept of civilian control in science and technology" as explained below.

At present, though civilian control is an extremely important principle to keep the military under check, it also has a drawback that the quality of military judgment would deteriorate since the supreme commander is not an expert in military affairs. Even here, "expertise" would come into play. Therefore, we can expect similar criticism for civilian control of science and technology. This may stir up the debate on whether "Amateurs from the field of humanities/social science with no known expertise in science and technology can regulate science and technology?" or "Wouldn't doing so lead to disastrous consequences?" Nevertheless, I think that civilian control of science and technology is indispensable to modern society....

People from the humanities/social science field are expected to acquire scientific literacy, and likewise, people from the field of science and engineering are expected to acquire social literacy. It is only then they can have two-way communications, and civilian control can happen.

In Japan, since humanities/social science and science are differentiated and taught as separate streams from a very early stage such as high-school, students from the science stream miss out on the opportunity to study human psychology or social behavior in detail. Consequently, scientists do not have the experience to weigh the impacts of science on the society. Therefore, even if a "republic of trans-science" problem presents itself, there is a risk of it being misjudged as a "republic of science" problem, causing things that can finally spin out of control. Therefore, Kobayashi stated that "civilians" need to "control" the behavior of scientists provided people from a humanities/social science background acquire science literacy.

When faced with trans-science problems, Weinberg discussed the code of conduct of scientists from the viewpoint of scientists while Kobayashi discussed what citizens must do from the citizen's perspective.

4.1.4　What Is the True Nature of Science?

The idea of "civilian control of science and technology" has been accepted as a topic worthy of attention by many experts since the TEPCO Fukushima Daiichi nuclear accident.

However, I would like to disagree with this concept by Kobayashi (Yamaguchi, 2015b). Although this discussion seems "justified" at first glance, this idea can intimidate scientists and suppress the evolution of science when we consider the true nature of science.

So, what is the "true nature of science?"

In his book "*Butsurigaku to wa nandarou ka* (What is Physics)?" (Tomonaga, 1979), Sin-Itiro Tomonaga (1906–1979) defined physics as the "pursuit of laws related to various phenomena that occur in the natural world that we are surrounded by—however, mainly the laws pertaining to the abiotic matter—while seeking evidence in observed facts."

If we expand this definition, we can define science as "the pursuit the laws underlying various phenomena that occur in the natural world that we are surrounded by, while seeking evidence in observed facts." So, what does "pursuit of laws while seeking evidence in observed facts" mean? Let us look at this, using the innovation diagram explained in Chapter 3.

As explained earlier, there are two kinds of processes of human intellectual activities, namely the "knowledge creation" and the "knowledge embodiment" (value creation). Innovation as acts of reform that creates economic and social value is produced through these two chains of activities.

"Knowledge creation" does not pertain to "knowing what no one knows or seeing things no one has ever seen" but also means "making it exist which has never existed." We can refer to these intellectual activities pertaining to the "knowledge creation" based on "abduction" as "research." The area beneath the border drawn horizontally in the innovation diagram is "soil" or "night science."

"Knowledge embodiment" is the intellectual activity of connecting and integrating such newly created "knowledge" with knowledge or with existing technology and molding it into economic and social "value." We can also refer to this intellectual activity of "knowledge embodiment" based on "deduction" as "development." This is the "leaf bud" sprouting in the area above the border or "day science."

The process of creating knowledge is moving forward in true darkness without the help of any manuals or textbooks and relying just on one's tacit knowledge. On the other hand, above the soil, there is bright sunshine, and this is the world seen by the market or society. Citizens who carry out economic and social activities live on the soil and cannot see "beneath the soil."

4.1.5   Border Between Science and Trans-science

If we look at Kobayashi's theory from the perspective of the innovation diagram, we notice that not only does it give any consideration to "what is science?" but also that it clubs the two completely different intellectual activities of human beings, namely "knowledge creation" (science) and "knowledge embodiment" (technology) under one heading, namely "science and technology" to pursue the argument. The "science and technology" mentioned in this argument completely ignores true science or "night science."

If we reconsider the relationship between science and society after properly defining the border between science and trans-science, a "consensus conference" with citizens and scientists should appear completely different from the opposing view of "civilian control of science and technology."

Let us assume that "Beneath the soil = Science" and "Above the soil = Trans-science" in the innovation diagram. In short, the border between science and trans-science is the surface of the soil in Figs. 3.2, 3.5, and 3.7.

Below the soil is the world of "knowledge creation" where there is no social/economic "value." This is consistent with the principle that in value-neutral science, "truth, goodness, and beauty" are not questioned, and the newly created "knowledge" is a general asset shared by the entire humanity. On the other hand, above the soil is the world of social/economic "value" that is created by "deductively" giving shape to "knowledge" produced in this way.

In this manner, defining the border between science and trans-science, we can clear up many misunderstandings and avoid unnecessary confusion.

Firstly, it can prevent "scientists" from getting intimidated by the concept of "civilian control of science and technology and the suppression of scientific evolution that may take place because of this fear."

In the world of "night science" beneath the soil, since the sprouting of various hypotheses and discoveries has not been converted to written form and continues to largely exist as tacit knowledge, scientists have to believe their own senses and imagination, and move ahead while groping in the dark.

If society understands that such a world exists, it will become possible to convince them about the significance of social investment. Since it is abduction that brings about paradigm disruptive innovation, the "night science" that should be driven by the curiosity of scientists will not be dismissed by society as "useless."

Secondly, scientists will be able to make an unbiased contribution to trans-science problems.

As Shibatani argued, in Japan, due to the lack of a clear definition of the border between science and trans-science, politicians could easily step into the world of science. Therefore, scientists often had no choice but to show solidarity with political parties.

However, if society acknowledges the social significance of "night science" and the politicians also ensure that they will not interfere with it, scientists will be able to actively participate in discussing trans-science problems.

Thirdly, it is possible to change the conflicting relationship between society and science and create a synergy between them.

As Weinberg asserted, if a trans-science problem appears, scientists are obliged to proactively reach out and share their views without any bias and favor. What ensures that scientists take the fair stance is "night science."

4.2 Two Symbolic Accidents

4.2.1 Fukuchiyama Train Accident That Was 100% Foreseeable

Based on the above discussion, let us re-examine two case studies of trans-science problems in recent years.

The first case is the Fukuchiyama train accident that was caused by JR West (West Japan Railway Company) in Amagasaki City, Hyogo Prefecture on 25th April, 2005, the worst accident in JR's history. As shown in Fig. 4.1, the rapid commuter service that had just left Itami station overturned while turning right at a curve between Tsukaguchi and Amagasaki stations, instantly killing 107 people and leaving 562 people injured.

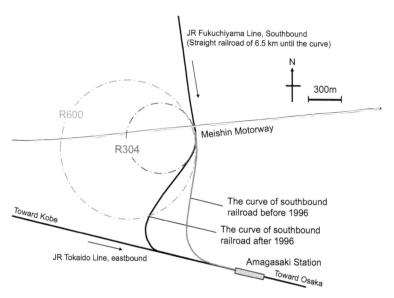

Figure 4.1 Map of the scene where the JR Fukuchiyama Train Accident happened in April 25, 2005. The radius of the curve where the train overturned was originally 600 m (gray curve). In 1996, JR changed the radius to 304 m (black curve). Adapted from Fig. 3.4 of Yamaguchi (2007).

The reason for the accident has been proved in the book "*JR Fukuchiyamasen jiko no honshitsu* (Root for the JR Fukuchiyama

Train Incident)" (Yamaguchi, 2007), that is, JR West did not make a physical estimate of the "overturn speed limit" when they changed the radius of the curved track from 600 meters to 304 meters in December 1996 and neglected the installation of the ATS (automatic train stop) device.

If we calculate the overturn speed limit using physics, as shown in Fig. 4.2, given that the railway curve radius is 304 meters, the overturn speed limit will always be 120 km/h or less with passengers more than 7. For the 93 passengers aboard the train (first car) on the day of the accident, the overturn speed limit was 106 km/h, and if the speed exceeded this limit, overturn would certainly occur with a probability of 1 (in other words, with a predictability of 100%).

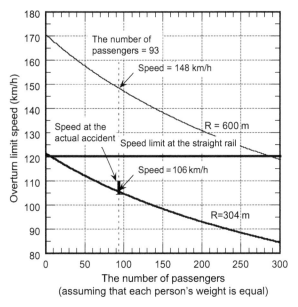

Figure 4.2 Overturn limit speed calculated by the overturn equation as a function of the number of passengers when the radius of the curve is 304 m and 600 m. Here, I assumed that each person's weight is equal. Adapted from Fig. 3.3 of Yamaguchi (2007).

On the other hand, when the radius of the curve remains 600 meters, the overturn speed limit will be always 120 km/h or above with passengers less than 208. For the 93 passengers aboard the train on the day of the accident, since the overturn speed limit was 148 km/h, even if the train had entered this right curve at 120 km/h, the probability of overturn was zero.

The speed limit for the right curve at the accident site was 70 km/h, and the speed limit for the straight track extending for 6.5 km before leading to the curve was set as 120 km/h. The 3 minutes 15 seconds required to cover the distance of the straight track is more than sufficient time if the driver of the train is no longer able to make a sound decision due to diminished capacity or disorientation. So, JR West which decided to change the design of the rail track should have changed it in such a way that the train does not overturn even if the driver enters the track at 120 km/h without decelerating.

In short, this overturning accident did not happen by chance and could have occurred with other drivers who were not told about the overturn speed limit. When the design of the curve was changed at the site, it was almost 100% foreseeable that this would happen in future. Nevertheless, neither did JR West take any precautions nor did they even calculate the overturn speed limit as shown in Fig. 4.2.

However, the mass media pursued only indirect and other factors in the background such as punitive management culture as the reason for the "driver's mistake." This was not what actually caused the accident.

From a scientific point of view, the most basic aspect of a technical enterprise in designing a railway track is to ensure that no serious accident occurs even when the driver has a temporary lapse of judgment. The real reason behind the accident is evidently the incorrect design of the railway track and negligence because the management did not know technology or in other words, it was a mistake of "technology management."

4.2.2 A Judicial System That Does Not Dismiss Science

However, the final report released by the Ministry of Land, Infrastructure and Transport's Aircraft and Railway Accidents Investigation Commission (Chairman Norihiro Goto) in June 2007 concluded that "the driver applied the brakes too late" and placed the entire responsibility of the accident on the driver.

Not only this, regarding the delay in the driver applying brakes, the report made a presumption that "since the driver was paying special attention to the communication between the conductor and transport commander, the driver may have been distracted and caused the accident." It also assumed that the "the driver was fit enough to make a sound decision" without any evidence. They did not hold the management responsible and instead pointed out that there was a delay in installing ATS and replacing it for the sharp curve near the site.

Following this, a number of criminal trials took place, and in the first criminal trial against Masao Yamazaki, the general manager of the railway during the time when changes were made in the route design, who was charged with professional negligence resulting in injury or death, the Kobe District Court returned the below "not guilty" verdict in January 2012:

> The driver entered the curve at a speed that exceeded the overturn limit speed, causing the train to overturn, leading to loss of life and casualties of traveling passengers. It is hard to deny that such things are within the foreseeable range as "for some reason" or "could happen someday."
>
> However, since the circumstances leading to the train approaching the curve beyond the overturn speed limit are vague, the possibility of the outcome is also not specific. In that sense, if it is predictable that the result occurrence is within the foreseeable range, then there is no big difference from the sense of fear, so the prediction of the outcome is not easy, and it must be said that the extent of predictability is also considerably low.

As you can see, this judgment dismisses science out of hand. Science correctly estimated the overturn limit speed. Therefore, it was easy to make a quantitative prediction. Moreover, the actual overturn speed matched the overturn limit speed that was determined theoretically (106 km/h). Science used in this calculation was high school level physics, which all the railway operators and managers ought to have known.

Then, why did the driver suffer a loss of judgment at the straight track before entering the curve? Why did the managers interfere with the installation of the ATS on the curve (given that it was installed in the train)? Why did not the managers know the overturn speed limit? The trial absolutely does not make any effort to answer these three questions using evidence.

Although Yamazaki was indicted without being held in custody, he was acquitted of these charges. Although the prosecution panel comprising citizens forcibly indicted all past three presidents of JR West twice, based on a vote that they "Just be indicted of the charges," they were eventually acquitted of all charges.

Though this case should ideally be a trans-science issue, it is a typical example of how the scientists and judicial authorities did not at all share their knowledge with each other or "ask questions of science."

4.2.3   Why Was Seawater Not Injected? Fukushima Daiichi Nuclear Accident

The second case is the Fukushima Daiichi Nuclear Power Plant accident that occurred between March 11 and 14, 2011. Let us look at the accident process first.

The technology of nuclear reactors ultimately boils down to the question of how the reactor core (fuel rod) in the pressure vessel is cooled down. At the Fukushima Daiichi Nuclear Power plant, the AC power and pumps to take in seawater were damaged due to the tsunami, and the core could no longer be cooled. The Emergency Core Cooling System (ECCS) that operates in such situations also stopped working because the emergency power supply was also shut down due to the tsunami.

Even if such a situation arises, as the "last resort" during an emergency at the nuclear power plant, the "IC" (isolation condenser) is installed in Unit 1 while the more evolved version of IC called "RCIC" (reactor core isolation cooling system) is installed in Units 2 and 3.

These should continue to operate even if there is complete power shutdown and cooling down of the core. In fact, as shown in Fig. 4.3, if we read the data on the reactor water levels sent from the accident site every few minutes, we can see that even after the ECCS shut down, Unit 3 RCIC (and high-pressure injection system, HPCI) for 35 hours and 39 minutes, and Unit 2 RCIC for 70 hours and 36 minutes continued cooling the core (Yamaguchi, 2012).

(a)

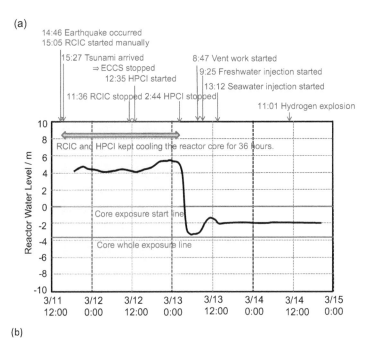

(b)

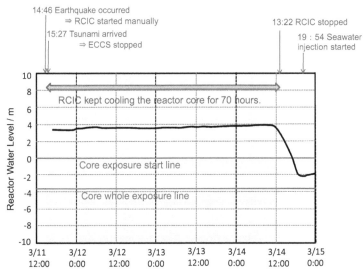

Figure 4.3 Time evolution of the reactor water levels for (a) unit 3 and (b) unit 2 of Fukushima Daiichi Nuclear Power Plant. Here, the gray area indicates the time region when RCIC and HPCI worked and kept cooling the reactor core. Adapted from Fig. 1.5 of Yamaguchi (2012).

In other words, the nuclear reactor did not go out of control soon after the ECCS shut down, so if some action had been taken during the time the "last resort" was being implemented, it would not have been impossible to control. At least for Units 2 and 3, it was possible to open the vents, a manual valve attached to the primary containment vessel (PCV) when the situation was still controllable so as to lower the pressure of PCV and inject seawater.

However, Ichiro Takekuro (former executive vice president cum managing director and general manager of Nuclear Power & Plant Siting Division), fellow of Tokyo Electric Power Co. Inc, who was TEPCO's management representative to the prime minister and his cabinet complied with the directions of Chairman Tsunehisa Katsumata and President Masataka Shimizu and stubbornly refused to inject seawater. It was only when the "last resort" of Unit 2 and Unit 3 also failed and core meltdown occurred that they decided to eventually inject seawater.

As long as the reactor water level is positive and the fuel rod is immersed in water, the reactor core does not melt. Thus, even if the vent is opened, water vapor containing only a very small amount of radioactive substance is released into the atmosphere. However, if the vent is opened after the reactor water level becomes negative and a part of the reactor core comes out of the water, a large amount of radioactive substance is released, causing serious radiation damage.

Moreover, if the reactor water level becomes negative, the reactor goes beyond human capability and becomes uncontrollable. This is when we say it has exceeded "physics limits." This was observed in the nuclear power plant accident of TEPCO.

On 13th March, when the reactor water level of both Unit 2 and Unit 3 was positive, the radiation damage would have been contained to only Unit 1 (1/6th of the current situation) if the vent had been opened to release pressure and seawater had been injected promptly by 3 am.

In addition, as shown in Fig. 4.3 (a), after the HPCI suddenly began to operate in Unit 3, the pressure inside the pressure vessel fell to less than 10 atm. Therefore, seawater could have been injected using a high-pressure fire engine without venting. Why was not seawater injected at this point? Why did they not even try?

In order to find out why the accident happened, the government, the Diet, and the TEPCO Accident Investigation Committee were set up to investigate the accident. However, both the government and TEPCO are affected parties, and the National Assembly Accident Investigation Committee is also a so-called official organization. In order to objectively analyze accidents and share unbiased facts with the citizens, the investigation needs to be done purely by grass-roots organizations.

So, the "Fukushima Project Committee" chaired by me was set up to study the true nature of the accident and make recommendations for the future from a third-party standpoint. Members from several fields like law, nuclear reactor engineering, energy problems, and technology ethics including the principal founder and former vice president of Matsushita Electric Industrial Co. (current Panasonic) Hiroyuki Mizuno came together to form this committee.

We analyzed the process leading up to the accident from the published data and published it in the *"Fukushima Repōto—Genpatsu Jiko No Honshitsu"* (Fukushima Report—The True Nature of Nuclear Accident) (Fukushima Project Committee, 2012).

4.2.4 Fear of the Nuclear Reactor Being Decommissioned

In all the three nuclear reactors, the "last resort" was being used, and it continued to cool the nuclear reactor core. All the necessary information for decision making was available with the management team at that time. It was 100% foreseeable that the reactor would become uncontrollable if the judgment was delayed.

However, it took 28 hours for Unit 1, 46 hours for Unit 3, and 77 hours for Unit 2 between RCIC (Unit 1 IC) operation and seawater injection. For Unit 1, though we can imagine that the site staff including Masao Yoshida, the plant chief did their best in the situation, they were hesitant to take the decision to inject seawater for an abnormally long period of time for Units 2 and 3.

Why on earth did the management team of TEPCO wait for the reactor to spin out of control and then take the decision to inject seawater?

I interviewed Yasushi Hibino (vice president of the Japan Advanced Institute of Science and Technology at the time), who was advising Prime Minister Naoto Kan and his cabinet on the night after the accident, so I was told by Hibino what happened at the prime minister's office at that time.

The topmost priority for the reactor technology management is to ensure that the reactor water levels do not become negative. Therefore, the prime minister repeatedly asked Ichiro Takekuro, TEPCO fellow, as well as Nobuaki Terasaka, director general, Nuclear and Industrial Safety Agency (NISA), and Haruki Madarame, chairman of the Nuclear Safety Commission, "for Units 2 and 3, shouldn't we immediately open the vent to release the pressure and begin injecting seawater?"

When Hibino also received the response that, "There is no danger from injecting seawater" from TEPCO (Nuclear Quality & Safety Management Department), he urged them, questioning, "Why didn't you open the vent and inject seawater as soon as possible? (While the RCIC is still running and the reactor water level is being maintained positive)."

However, Mr. Takekuro, knowingly and willingly refused to do so.

What was the reason? TEPCO's explanation that, "We can release large amounts of energy by opening the vent when the temperature and pressure of the reactor pressure vessel (RPV) is the highest, so we will release the vent after the RCIC stops, and inject seawater" given to Prime Minister Kan and Hibino has no rational basis in physics. The actual reason is probably "if seawater is injected, the reactor will be decommissioned, and this will cause significant financial losses."

4.2.5   Concealing Management Responsibility

"The top priority of the reactor technology management is to ensure that the reactor does not surpass "physics limits' and spin out of control, that is, making sure that the reactor water levels do not become negative." Therefore, the procedure "when RCIC begins to operate as the last resort after ECCS fails to work, the vent must be released as soon as possible, and while the water levels are positive, freshwater should be injected. Meanwhile, the required arrangements to inject seawater need to be made, and it needs to be injected instantly" must be strictly followed. This is the "Normally-OFF" principle.

Here, "Normally-OFF" is the principle in which the "technical system is designed to turn off when the power supply is stopped (normal)," and this is the basic principle to prevent the system from running out of control in unforeseen circumstances.

This principle can also be generalized as follows: "The technical system is designed to run such that it does not surpass physics limits and spin out of control in unforeseen circumstances." This universal principle is not only the fundamental principle of technology design but also the fundamental principle of technology management.

However, the nuclear reactor, in principle, does not satisfy the universal principle of "Normally-OFF." When it is cut off from the external power supply, RCIC or the last resort will work but only for a limited time. When the time limit runs out, the reactor water level will become negative, surpass the physics limit, and spin out of control. In other words, the nuclear reactor is based on the defective technology of "Normally-ON." In order to compensate for this defect and satisfy the "Normally-OFF" principle, the seawater needs to be injected when the RCIC is in the running condition.

It is interesting to know that the TEPCO management did not know even this basic principle. Therefore, they did not understand what the "physics limit" of the reactor was, which when exceeded would surpass human capability and cause the reactor to spin out of control. In short, the real reason behind this accident is not "technical fault" but "negligence of technology management."

Regarding the accident, the defects in the so-called risk management plan such as the construction of a nuclear plant just 11 meters above sea level and installation of most of the emergency power supply on the basement, were questioned.

Of course, this is gross negligence. However, when looking into the cause of the accident, turning our attention only to risk management, will make us misunderstand the essence of the problem. The essence lies not just in risk management but also in the negligence of technology management.

When the accident occurred and the nuclear reactor spun out of control, the engineers should have known what would happen later.

I heard that the former director of NISA wept bitterly and said, "It is all over." However, TEPCO managers did not know that the reactor would spin out of control if the reactor water levels became negative, nor did they even make an effort to understand this fact.

In spite of all this, they continue to conceal the discussion on management responsibility in the accident. The prosecution declared that it would have been "difficult for them to predict the earthquake and tsunami" and dropped all charges against the TEPCO management and government officials related to the accident. The committee for the inquest of the prosecution charged the three executives Tsunehisa Katsumata, chairman of TEPCO at the time of the accident, Sakae Muto, vice president, and Ichiro Takekuro, fellow with "neglecting to take appropriate measures despite receiving reports that a tsunami as high as 15 meters could occur." They were forcefully indicted in February 2016.

4.3 Why Is Scientific Thinking in Organizations Lost?

4.3.1 Who Are the Experts?

The lack of scientific literacy in both the organizations is the common factor for the cause of "negligence of technology management" in the JR Fukuchiyama train accident and the TEPCO nuclear accident. In order to ensure scientific thinking within a company, I believe that scientific experts should be included in the decision-making system.

So, who are the "experts"?

After the Fukushima Daiichi Nuclear Power Plant accident, the National Diet of Japan Fukushima Nuclear Accident Independent Investigation Commission was set up on May 28, 2012 to investigate the prime minister's interrogation of the mishap. Shuya Nomura (professor, Law School, Chuo University; lawyer) who got to hear the interview asserts that "Prime Minister Naoto Kan called Yasushi Hibino, who is not a nuclear expert to his office, and made him ask several questions to the Fukushima directors from a layman's perspective and caused unnecessary confusion at the site."

Yasushi Hibino was a qualified physicist. However, Nomura had a stereotypical notion that "anyone who is not an engineer or professor of nuclear engineering is no expert."

So, should people not qualified in nuclear power engineering step into this discussion? I would rather think that the problem lies in the fact that scientists and social scientists from disciplines other than nuclear power engineering have not joined the nuclear power plant administration to have an open discussion. However, the reality is very much the closed community, namely Japan's "nuclear village." In order to solve problems during emergencies, we should also involve physicists who are not part of this "village," so Nomura's argument does not scientifically have a point.

Nomura loudly exclaimed "Kan risk" during the interrogation. He said that "Though Prime Minister Kan is not an expert, he played expert, only making this issue more serious." The accident investigation report of the National Diet of Japan Fukushima Nuclear Accident Independent Investigation Commission that came out one month later accused Prime Minister Naoto Kan's administration of hampering operations at TEPCO as the disaster unfolded. And the drift of the argument presented by the National Diet of Japan Fukushima Nuclear Accident Independent Investigation Commission has become the one national consensus. But is this really the truth?

For Unit 2, RCIC shut down eventually on 14th March at 13:22. In a matter of hours, the reactor water level became negative, resulting in the meltdown. In the evening of that day, a crack suddenly appeared at the bottom of PCV. Seawater was injected from 19:54 without opening the vent and reducing the pressure inside the pressure vessel to 8 atm or less, but a large amount of radioactive material of high concentration from the core meltdown was released from the reactor through the crack. As a result, Unit 2 spun completely out of control, and the situation became so bad that it became impossible to even enter the building.

In the evening of March 14th, President Shimizu of TEPCO called Banri Kaieda, the then minister of Economy, Trade and Industry and told him, "I would like all the TEPCO's personnel to evacuate, leaving the nuclear reactor that has gone out of control as it is." However,

Prime Minister Kan dismissed TEPCO's "request to evacuate." After calling President Shimizu to the official residence, Prime Minister Kan said, "Do you know what will happen to Japan if you evacuate now?" He instantly traveled to the TEPCO head office and set up the emergency response office at the site.

In response to TEPCO's request to evacuate, it is said that the "experts" in the government and the Ministry of Economy, Trade and Industry concluded that there was "no choice but to evacuate." If TEPCO had evacuated completely from the accident site following the advice of these "experts," as the Atomic Energy Commission had foreseen it as a "contingency situation" at that time, Tokyo would have become uninhabitable today. Regarding the "evacuation," TEPCO insists that it meant that they would "leave the minimum required personnel for managing the situation at the site," but it has not been verified with facts that "they had insisted on this from the beginning and not in retrospect."

Kan and Hibino, who are not "experts" according to Nomura, urged the personnel concerned to "open the vent immediately and inject seawater," but the technology management representatives, TEPCO Fellow Takekuro and Vice President Sakae Muto, the "experts" according to Nomura, refused to oblige. These so-called "experts" were responsible for the accident.

The problem here is that even though Takekuro came from a science background (Faculty of Engineering, Department of Mechanical Engineering), why did he refuse to inject seawater?

After joining TEPCO, he pursued a career in the nuclear field, taking charge as director of Kashiwazaki-Kariwa Nuclear Power Site, vice director of Engineering Research Development Headquarters, and as the vice president and the general manager of Nuclear Power and Plant Siting Division, which is the topmost position in the nuclear field. Of course, he should have had a thorough knowledge of the finer aspects of technology such as the piping structure of a nuclear reactor. But it appears that he had no knowledge of the "physics limit" which is the core of the nuclear reactor.

Simply put, the problem presented by the nuclear accident does not stop at the segmentation of humanities/social science and science rather, it is a typical case where refusal to "knowledge cross-border" between different fields of science made it difficult to solve trans-science problems.

4.3.2 Nuclear Power Policy That Excludes Physicists

At the Fukushima Daiichi nuclear power plant, the emergency power supply was placed in the basement. During the tsunami, the power supply was submerged, leading to total power loss. The cause of such elementary "misplacement" can be traced back to the time when Japan introduced nuclear power. Here, we also encountered a similar problem with the "experts."

In the 1950s, the people involved in nuclear policy promotion in Japan were Matsutarō Shōriki, who owned Yomiuri Shimbun, Yasuhiro Nakasone, the first Chairman of the initial Atomic Energy Commission and member of the House of Representatives. Hideki Yukawa, the Nobel laureate in physics, was also invited to be a member of the Atomic Energy Commission.

However, Yukawa, who asserted the importance of basic research, opposed Shoriki's policy of early commercialization of nuclear power by "directly importing" it from the United States and resigned from the Commission in little more than a year. Subsequently, Yukawa backed out from the front line of nuclear power development, and there has been no involvement of physicists in Japan's nuclear policy promotion since then.

Thus, Japan's nuclear power policy came into existence as a "turnkey contract" with the General Electric Company (GE) of United States. This kind of contract means that Japan must accept all GE designs without incorporating Japanese science and technology.

Unit 1 of Fukushima Daiichi Nuclear Power Plant's decision to install all emergency power supplies in the basement was actually GE's design to guard against tornadoes, which occur frequently in the United States. GE could not visualize scenarios that are peculiar to Japan such as the arrival of the tsunami, and Japan also did not interfere with their design.

Units 2 and 3 were built by Hitachi and Toshiba, but the construction was based on GE's design, which the Japanese were not allowed to modify. The Japanese manufacturers, such as Toshiba, could modify the designs of only Units 5 and 6.

Therefore, a part of the emergency power supply of Units 5 and 6 was placed on the first floor in an air-conditioned space, so they remained safe during the tsunami. Also, the Onagawa power station of Tohoku Electric Power Co. in Miyagi prefecture, unlike TEPCO's Fukushima Daiichi Nuclear Power Plant (10 meters height above sea level), was built at a height of 15 meters above sea level, so it also remained safe during the tsunami.

If we go back into the past to understand the cause of the TEPCO nuclear accident, it is clear that Shoriki kept Yukawa completely out of the picture. If Shoriki had valued Yukawa's opinion as a physicist and entrusted him with the initiative, young physicists would have come together and designed the nuclear reactor more logically. Moreover, they would have also been more actively involved in the nuclear power administration. I believe, under such a scenario, this accident would have never occurred.

From 1958 to 1967, the nuclear engineering department was set up as a part of the faculty of engineering in the former seven imperial universities of Japan to train nuclear engineers. However, physicists like Hideki Yukawa and Sin-Itiro Tomonaga were not allowed to take the initiative. In this manner, a closed community called a "nuclear village," centered around nuclear power engineers and which completely distanced itself from physicists, grew like a monster.

4.3.3   JR's Exclusion of Scientists

The process of excluding scientists from the nuclear power administration led to the nuclear accident, and a similar situation was seen in JR West with the Fukuchiyama train accident.

Before shifting to JR, Japan National Railway (JNR) had all the competent scientists and engineers who designed the Shinkansen. Vibration engineering scientists, who had engaged in the design of fighter aircraft before World War II, entered the world of railway technology after the war. They designed the railway tracks in such a way that trains would definitely not overturn at each curve. In fact, they laid a test line in Hokkaido to conduct train overturning experiments and undertook research by repeating the experiments hundreds of times.

However, the privatization of the National Railway in 1987, which produced the Shinkansen, split the research institute of the headquarter as a non-profit foundation, the Railway Technical

Research Institute. At this point, the group of researchers in the National Railways were separated from JR and moved to the non-profit organization, and their existence was undermined.

New scientists of land transportation lost their link with the site and also the links with other disciplines through the site. On the other hand, JR West as a business division lost their scientific thinking capability by chasing away scientists who could make scientific investigations.

Even in the Fukuchiyama line where the accident occurred, the speed limit of the straight track extending over 6.5 kilometers during the period of JNR had been set at 100 km/h. As shown in Fig. 4.1, since the radius of the inbound track of the right curve at the accident site was 600 meters, the overturn speed limit was calculated to be 120 km/h or above, as shown in Fig. 4.2. Nonetheless, to ensure further safety, the speed limit had been further restrained.

In other words, "even if the driver suffers a loss of judgment due to diminished capacity or disorientation within those 3 minutes it takes to cover the straight track, the design was such that the "Normally-OFF" principle is upheld, and the speed of the train does not exceed the "physics limit."

However, under competition from private railways like Hankyū, JR West raised the speed limit to 120 km/h in the 1990s without any scientific consideration. Moreover, on the right curve of the track at the accident site, where the speed limit drops to 70 km/h, they neglected to install the speed control system (ATS).

What is strange is that there were about 25,800 employees in JR West at the time. Among the engineers who knew physics and the drivers, there must have been employees who would have noticed that the overturn speed limit seemed to be rather low. However, this feedback did not reach the JR West management team who had lost their scientific thinking capability, and even if it had reached them, it would have probably been ignored.

In both the JR accident and the nuclear accident, prosecutors and judges conducted investigations and proceedings on the premise that management has no scientific thinking capability. Science cannot be discussed in the world of the judiciary. In this situation, if people from the humanities/social science who do not know the meaning of "physics limit" and who have no science literacy are made representatives of such organizations, all such accidents caused would be exempt from their responsibility. Consequently, even when major accidents strike Japan, we are stuck in an odd situation where nobody assumes responsibility.

4.3.4 Incorporating Science in Organizations

All technologies inevitably have "physics limits." This determines the border between the "controllable" technology dimension and the "uncontrollable" technology dimension. If this limit is exceeded, it will surpass human capability and will cause the train to overturn, the plane to crash, the ship to sink, and the reactor to spin out of control.

The trans-science problems of these two accidents have not been solved to this day.

And the lack of scientific literacy is not an issue specific to TEPCO or JR West alone, but a universal problem that is deeply rooted in Japanese society. This means that similar accidents can happen again, and such disasters, due to the lack of scientific thinking, can occur in other industries dealing with technology such as the pharmaceutical or energy industry.

Why did the "negligence of technology management" effect both the accidents? If we look into the background of these cases, we realize that it is because TEPCO and JR West were organizations that did not require innovation.

High-tech companies in the midst of fierce international competition cannot survive unless they achieve breakthroughs. On the other hand, TEPCO and JR West are actually oligopolies or monopolies, and they hardly need innovation to survive.

Under these circumstances, there is no choice but to evaluate employees on a penalty point system. As far as risk management in a world based on penalty points is concerned, there is a tendency to look at "how to stay away from risks" rather than "how to minimize damage" in unforeseen circumstances. This culture inevitably takes away creativity and imagination from the organization.

These accidents exposed the lack of technology management capabilities in monopolistic companies that do not require innovation. These accidents were a warning to Japanese society that if "there are no more breakthroughs, the Japanese industrial system will no longer be accepted in the world."

In Japanese society, the various branches of academic disciplines operate in silos, and there is a deep-rooted tendency to treat transcending academic disciplines as taboo. People from the humanities/social science and science do not interfere in each other's fields, and each has its own special culture.

In other words, in addition to the absence of scientific experts in the organization's decision-making system, we can say that the lack of a network for collaboration across academic disciplines was the structural factor that caused the unfortunate accidents of TEPCO and JR Fukuchiyama.

To overcome the "technology management error," the humanities and social science must acquire science and technology literacy while science must absorb social science literacy. Moreover, there is a requirement in the concept of solving problems to transcend the borders of academic disciplines and absorb transilience to produce innovation that creates new value. Let us examine this concept more specifically in the next chapter.

Chapter 5

Social System That Produces Innovation

5.1 Reconstructing the Fields of Resonance

5.1.1 After World War II, Japan Tried to Create a Society That Does Not Take Risks

To understand why the drive for innovation ceased to exist in Japan since the beginning of the 21st century, it is important to look at the institutional design of post-war Japan.

I took up three policies that had already suffered huge setbacks in post-war Japan after the 1990s and made a detailed analysis of their causes (Yamaguchi, 2006a). These three policies, namely the agricultural policy, measures for birth rate decline and the aging population, and regional promotional policy are pressing issues even today. As a result, I found that there are common problems underlying these three seemingly unrelated failures.

It appears symbolically in the agricultural policy. After the war, the agricultural policy aimed at protecting farmers and declared agricultural land as a property instead of a productive resource. In other words, the market principle could not be applied to determine the price of farmland, which eventually led to a vicious cycle that lasted for decades with part-time farmers, who escaped the property

Innovation Crisis: Successes, Pitfalls, and Solutions in Japan
Eiichi Yamaguchi
Copyright © 2019 Pan Stanford Publishing Pte. Ltd.
ISBN 978-981-4774-97-0 (Hardcover), 978-0-429-44862-1 (eBook)
www.panstanford.com

tax payment, amassing more wealth than full-time farmers whose main livelihood was from their farms.

However, full-time farmers are exemplary entrepreneurs in the first place. Thoughtlessly protecting farmers with no managerial ability will only hamper the revival of the industry. Depriving equal opportunity to newcomers and going against productivity improvements will lead to a reduced population of farmers as they age and cause serious problems of low food self-sufficiency rates.

Of course, before the war, the policy of protecting farmers from the weather and economic risks to ensure that the misery that often struck the farmers does not repeat itself had brought about reliable outcomes for about half a century after the war. One could argue that this brought about "balanced development of land" in Japan, as agricultural land did not go waste, and industries were also promoted at the same time.

However, this postwar policy aimed at creating "a society where people can lead a happy life without taking any risks" resulted in a society that continued to remain as a vested interest for a long period of time, even after Japanese society matured and became a post-industrial society. This robbed the society of its courage and spirit to take on risks at various instances in the industry.

For example, the lifetime employment and the seniority employment system deprived the Japanese of their entrepreneurial spirit for over fifty years. After joining a company, though you could earn only a low wage at first, if you continue doing the work that you have been assigned for the time being, these wages would be enough to ensure a stable and secure life. It was much easier if you stuck on to the company as you would be sheltered from the storms of life.

Even in companies, if the scientists or engineers were transferred to production management or sales, there were hardly any employees who quit their jobs to establish start-ups with the aspiration of achieving their full potential.

The Japanese mindset was molded by such a post-war institutional design which continued to support the framework of the Japanese society for a long time. So, one of the Japanese goals to achieve success was to join a stable, large-sized corporation or government office and spend the rest of their lives in that organization. On the other hand, independent businesses that supported the region gradually declined, and local innovators began to disappear.

In other words, it is the institutional design of compensating risks arising from an uncertain future to the extent possible with taxes that robbed the Japanese of their courage and their ability to take on risks. As a result, the competitiveness of the key industries such as agriculture was hit from the inside, and independent businesses and entrepreneurial spirit began to decline and wither.

5.1.2   The Rules of Competition Have Changed

On the other hand, Japan has created a unique mechanism of technological innovation after the war. If we trace this process, large enterprises that prioritized industrial recovery in order to somehow catch up with international standards, began to set up research laboratories one after another in the 1950s. By 1960s, the number of researchers within enterprises was more than the number of university researchers, and talented researchers from the Faculty of Science and Engineering unanimously chose to join enterprise research laboratories instead of continuing research in their universities.

However, initially, these laboratories were places for technological improvement aimed at commercialization. The laboratory would often be located close to the production site, and there was no distinction between scientists and engineers.

In the 1980s, the ambitious research desire of large enterprises allowed them to eventually broaden the scope of basic research into science. This resulted in many technological innovations such as the high electron mobility transistor (HEMT) and blue LED originating in Japan. Japanese corporate research laboratories became an engine of technological innovation with few parallels around the world.

To reach this stage, a Japanese-style model where large enterprises became "self-sufficient" was adopted, and an unidirectional "linear model" of basic research (scientific research) $\rightarrow$ applied research $\rightarrow$ development of technology $\rightarrow$ commercialization was put into practice. The entire industry that was supposed to engage in positive rivalry became a "convoy" and worked under a system where they safeguarded each other from risks.

They believed that it was definitely better to create a society where one can live a stable and secure life without taking any risks. In a newly industrialized society, where the working population and wages were on the rise, such a risk-free society worked well.

However, as soon as Japan joined the league of developed nations, the "rules of competition in a global society" changed. From the late 1980s, the United States became essentially a high risk-high return society, where people who dared to take risks were respected.

Since the 1970s, there have been in-depth discussions in the United States on the new framework of innovation systems. In a post-industrial society, it is innovation that brings about sustainable economic growth.

So, when the "linear model" no longer holds good, where can innovation come from? As in the case in the semiconductor industry that had its roots in the invention of the transistor in 1947, it came from the next wave of start-ups established by people who popped out of the sources of innovation, such as the research labs of large corporations and universities. Unless innovators venture out and create new companies, the revitalization of industries producing innovation will not take place.

In this manner, the United States introduced a system to revive industries and enforced it by introducing laws one after another in quick succession.

Japan, on the other hand, lagged behind the rest of the world as far as start-ups were concerned. Research where you are unaware whether or not commercialization can happen is the riskiest. After plunging into global recession, large enterprises that were involved in international competition could no longer spare investment for basic research.

In such a scenario, it is necessary to change the traditional innovation model and "convert the results of research into economic and social value, and take on the challenge of innovation" and adopt a model like in the United States, where "people who took risks earn returns."

However, the idea that "start-ups are the engines of innovation" was not ingrained in Japanese culture. This was despite the fact that start-up companies such as Matsushita Electric (current Panasonic), Sony, and Honda led Japan in the past, and start-ups such as Nichia Corporation, Samsung of Korea, and TSMC of Taiwan were becoming the new industry leaders in recent times.

The Ministry of Economy, Trade and Industry and other central government agencies continued to distribute subsidies to large enterprises and their associations even in the field of high-tech industry. This consequently prevented new start-ups from coming up and the renewal of related industries.

I think that the root cause can be traced to the ideology on the basis of which the nation was rebuilt after the war, namely "building a society where you can lead a stable and secure life without taking risks." We must quickly create a new innovation generation system relevant to our times. Japan can afford no further delay in adopting the design of a social system which values the first penguin that springs out of the herd and dares to jump out into the rough seas. However, the problem is the methodology that is required for this purpose.

5.1.3   Reconstructing the Fields of Resonance in Universities, Industries, and Society

For Japanese companies to restore the innovation model based on "knowledge creation," which is the essence of science, the crucial task is to "transcend the borders of knowledge" while overcoming the barriers between academic disciplines and teach people who can perform "transilience."

This depends on whether we can rebuild the broken "fields of resonance."

I would like to consider the possibility of reconstruction of the fields of resonance mentioned in Chapter 3 at three levels, namely university, industries, and society.

The fields of resonance that produces "paradigm disruptive innovation" operate at the intersection of the "knowledge creation" vector and "knowledge embodiment" vector in the "soil" depicted in the innovation diagram. It acts as a platform where tacit knowledge could be spread from one person to another. The creators of the fields of resonance are people who have a thirst to create new technologies, taking inspiration from technologies and paradigms of completely different fields.

These are none other than start-ups that can provide a system to give shape to "abductive" knowledge. However, unlike the time when the fields of resonance existed in large enterprises in the past, since scientific research can be carried out only in universities and public research institutions in general, the place for "knowledge creation" and the place for "knowledge embodiment" have been physically separated. There are no fields of resonance anywhere where these two can happen concurrently.

The fields of resonance cannot be created through traditional industry-academia collaboration, where large enterprises request universities and obtain the outcomes of research. In both the quantum mechanics industry and the pharmaceutical industry, the argument given by the management to scale down research activities is that "if we have a collaborative model where research is carried out at the universities, and development is done by industries, there should no problem at all." However, the industry and academia in Japan did not see eye to eye from the beginning and disagree with each other even today.

This is because they do not understand the completely different human intellectual activities for conducting "research" and "development." They also do not make an attempt to create "fields" to create resonance between the ideas of people who dedicate their lifetime to these activities, based on their understanding of the difference between the two intellectual activities.

If the system is built based on the concept of trying to effectively create such fields of resonance, Japan should have acquired a new innovation model.

Eventually, it is necessary to give shape to the extraordinary "knowledge" created at the university and make use of it as innovation. Considering the manner in which the industry-academia collaborates today, this is absolutely impossible. People who can achieve this are team members who have researched this and understood it by experiencing the spirit behind this "knowledge."

Therefore, the most reliable way to create fields of resonance is when the team members, who were engaged in this research at the university, convert the "knowledge creation" into "value creation" as a member of this "field."

However, as we have seen so far, there is no motive, mechanism, and concept where these young, obscure scientists can be shaped into innovators of start-ups.

5.1.4   The Secret of the Cambridge Phenomenon

As a symbolic example of creating fields of resonance, let me introduce Britain's "Cambridge Phenomenon," where a number of leading university-initiated start-ups were established one after the other. I studied the "Cambridge Phenomenon" for a year from 2008 onwards when I was a visiting fellow at the Clare Hall, The University of Cambridge. Every summer after that, I stayed at The University of Cambridge for more than a month and continued my observations there.

The University of Cambridge has strengthened its collaboration with industry since the 1970s. It started with the submission to the government in 1969 of the "Mott report" by Sir Nevill Mott (1905–1996), professor and recipient of the Nobel Prize in physics, and of Cavendish Laboratory (Physics Department) at The University of Cambridge.

In this report, Mott talked about the collaboration between the university and industry to promote a site for science-based industries, or in other words, recommended that the university should be more involved with industrial society. At that time, considering that in Japan it was a time when students and universities would turn a cold shoulder if someone proposed industry-academia collaboration, this assertion that was full of foresight accurately predicted the dynamism that was witnessed in Cambridge ten years later.

In response to this report, the Trinity College, The University of Cambridge built a "science park" spread across 11 hectares on its campus in 1970, invited high-tech companies, and promoted joint research with the university.

Until then, unlike Oxford, the automobile industry was not being developed in Cambridge city. However, taking advantage of the "Mott report," high-tech companies started concentrating organically, and by 1990, more than 800 high-tech companies accumulated in and around Cambridge city. This was not due to the government or local authority's efforts to attract these companies, but it was completely a self-organized cluster.

The person who made the Cambridge Phenomenon a legend was Herman Hauser (1948–present), who was a graduate student at the Cavendish Laboratory. He did not follow the routine path of taking a doctorate in physics and becoming a physics researcher and instead founded Acorn Computers Ltd, a computer manufacturing company in 1978.

Hauser, who made millions from this venture, was later involved in commissioning another start-up called ARM, which I have briefly discussed in Chapter 3. I have already mentioned previously that the ARM-designed microprocessor became the standard chip for all mobile phones.

Well, if we trace the source of these two start-ups, Acorn and ARM, we come across the 31 colleges in Cambridge University. These colleges offered lodging along with breakfast or lunch besides dinner. In addition, you can receive private lessons from a faculty (fellow) of Cambridge University. These colleges are responsible for the entrance examination, and the University is responsible for the graduate certification.

A Japanese organization that closely resembles this system is probably the world of sumo stables. The world of martial arts also has a similar structure, where the "entry is in each dojo (training hall), and the tournament is in each martial arts organization." Many of the start-up companies that gave rise to the Cambridge phenomenon, including ARM, are companies set up together by friends "who ate off the same trencher."

Even at Clare Hall, the colleges where I stayed, the fellows and students ate "British food" together. There is an implicit rule here to occupy the seat in the order in which you have arrived. Therefore, it is not preferred that you sit with only friends or with people you are acquainted with. Then the network becomes inflexible.

When fellows from the humanities sat next to me, I was required to discuss philosophy or history with them for about two hours, including the coffee or drinks break at the salon. Since I was expected to represent Japan there, I decided to study history and philosophy so that I could have meaningful conversations on these topics. This experience became a huge stepping stone for me to propose "transcending the borders of knowledge."

This was a field of resonance where people of different age groups, gender, nationalities, and academic disciplines came together and identified with each other's values and academic disciplines. The resonance of knowledge and spirit eventually led to new innovations that the world had not seen.

5.1.5 Potential of "College"

Actually, I studied the Cambridge Phenomenon as I was requested by Masao Horiba, founder of the world-class high-tech manufacturer Horiba, Ltd., to "look into the secret of the one and only Cambridge Phenomenon" before I went to England.

My response to this was that after returning to Japan, I proposed to Horiba to set up a unique institution and Masao Horiba immediately "established a Kyoto college" in Kyoto city. Horiba took this proposal to Hiroshi Matsumoto, president of Kyoto University at the time.

Matsumoto comprehended this idea in his own way, and he set up the new Graduate School of Advanced Integrated Studies in Human Survivability (GSAIS), Kyoto University in 2013, where I currently work. This graduate school is conducting a five-year integrated doctoral course for creating global leaders who can address social problems such as environmental issues, human rights issues and innovation on a global scale, by helping them develop practical skills through resonance between humanities/social science and science, and transilience capabilities by "knowledge cross-border."

In the 21st century industrial society, the teams that have come together with fields of resonance as the core, based on a specific vision and purpose, will play a key role in the industry and rebuild the innovation chain.

Therefore, in order to speed up this process, for example, the university faculty members can become producers or intermediaries linking these activities. When university faculty creates a network of all these fields of resonance, it can also be actively involved in acquiring the "knowledge that can be embodied."

Moreover, the fields of resonance, including large enterprises, will soon discover a new collaboration network to enable them to harmonize their "knowledge that can be embodied" and build an unprecedented value chain to escape the "mountain climbing trap."

In fact, in Japan, from the 1970s to the early 1990s, there was an attempt to create fields of resonance across enterprises in the world of semiconductors and devices. There was a time when scientists and engineers of enterprises who identified with each other at academic meetings, overcame organizational borders to freely discuss unpublished tacit knowledge and enhanced Japanese technical capabilities.

I recall the experience even to this day. One engineer of a particular company in a certain field of resonance that surpassed organizational barriers narrated his experience, "Since the yield of HEMT is not increasing, we analyzed and studied the atoms and found impurity elements with specific atomic weight that are usually not found in semiconductor interfaces."

When he said this, the scientists and engineers from other enterprises also commented "Oh, you guys also? Actually, we did too. And we didn't know what to do." Three days and three nights following this, they brainstormed over the causes unrestricted by the limitations of their enterprises. In this manner, they found a solution to this unknown issue, and the performance of Japan's HEMT surpassed the rest of the world.

However, since the mid-1990s, such fields of resonance collapsed due to pressure from the management of these enterprises, and the fields of resonance that were formed within these organizations were lost without their importance being truly understood. Most of the aforementioned scientists and engineers were either transferred to the business divisions of their companies or quit their enterprises and simply vanished.

5.2   Institutional Reform of Universities and Industries

5.2.1   Nurturing Innovation Sommeliers

As I have mentioned so far, after the end of the "central research laboratory model" that occurred in the late 1990s, Japan has been drifting without being able to find a suitable innovation model for the 21st century that could replace the old model.

In order to rescue innovators from the scattered boat that is drifting, Japan needs to immediately put in place a system that embodies the American SBIR concept.

For this purpose, it is essential to establish a professional scientific administrative system that has so far never existed in Japan. As it happens in Japan today, people who do not have a doctorate and who have never taken risks associated with "abduction" or "transilience" should not be involved in the scientific administration anymore.

It is important to develop people whom I call "innovation sommeliers" who can conceive the future industry, understand the overall structure of innovation, and create "abduction" and "transilience" scenarios as part of university education.

However, people who believe that innovation only refers to the value creation through "knowledge embodiment" (Type 0) have not developed the skill of "abduction." Moreover, since both natural science and social science had a negative outlook on "knowledge cross-border," university education does not teach us "transilience" or freely surpassing the barriers of humanities/social science and science. As a result, people who explored only one field at a Japanese graduate school by digging in the direction of the soil knew only that field and could become researchers having skills with only limited applicability.

Then, going forward, how can we establish a system in Japan that teaches us how to perform "abduction" through "transilience"?

To achieve this goal, the first step is to set up a new graduate school. In this graduate school, social scientists and scientists would identify with each other on a daily basis to get to the bottom of things. On the other hand, scientists can learn the methodologies of social scientists or humanities scientists and have discussions in a common language. The goal is to create a new transdisciplinary field and "build a grand design" where science is properly incorporated into society by forming a circle according to the core academic disciplines in an "academic landscape" (Fig. 2.8).

However, this in no way means putting, say for example, econophysics in that area. This is because econophysics is just one field that "addresses economic phenomena using mathematical models." Instead, physicists fundamentally should try to understand next-generation semiconductors based on quantum mechanics, and at the same time, they should conceive a future vision of the semiconductor industry by studying economics and business administration.

In this way, the important thing is to master the ability to venture across two or more fields; for example, in the process of taking a doctorate, you can study two or more related academic disciplines. This is the essence of "abduction" and "transilience."

In this way, even when social issues appear, one "transcends the borders of knowledge" and utilizes the wisdom gained from multiple academic disciplines to verbalize the issues and find solutions.

For example, when an issue that concerns various academic disciplines such as the nuclear accident occurs, we analyze the problem and make it explicit knowledge through the resonance of natural science and social science to look at a solution. The new academic discipline should make this possible.

5.2.2 New Graduate School Design

What should the graduate school curriculum to develop innovation sommeliers be like? To obtain the answer, I would like to review the structure of the "academic landscape" with the 39 academic disciplines once again.

As mentioned in Chapter 2, the ten core academic disciplines existing at the center of the circle (mathematics, physics, informatics, chemistry, life sciences, psychology, philosophy, economics, law, and environmental studies), while having close interactions with each other, also have strong interactions with either of the nearest five clusters (engineering sciences, medical sciences, humanity and social sciences, management sciences, and geosciences).

In the case of those scientists who were accepted under the SBIR program in the United States and became entrepreneurs, a majority of them were from core academic disciplines such as physics, chemistry, and life sciences. There were also entrepreneurs with a doctorate in the core academic discipline of humanities.

Therefore, it can be said that the new graduate school that develops innovation sommeliers should be based on the following curriculum design concept:

First of all, this graduate school student should master the ten core fields of study as broadly as possible. Even if he does not understand the techniques of each of these academic disciplines, it is fully possible to understand the fundamentals and study them to the level that is required to speak a common language.

Secondly, this graduate school should be a lodging style school where students "eat off the same trencher." in the same way as in the 31 colleges of the University of Cambridge, where natural scientists, social scientists and humanities scientists exist in the same fields of resonance. In such a scenario, the students and the graduates are college mates who can complement each other in the ten core academic disciplines.

In addition, the school dormitory must emphasize diversity like in Cambridge College. It goes without saying that the vulnerable groups of society, including single mothers, should be given priority.

As mentioned above, college mates played an important role in both the start-ups Acorn and ARM that Hauser set up. Such lifelong companions who can create synergies between each other's existential desire for life are always required for integrating the network of innovators.

Thirdly, this graduate school student confronts various social issues related to human survival, acquires a doctor's degree, and internalizes "abduction." It is this experience of "making it exist which has never existed" that makes a breakthrough possible.

This "new transdisciplinary field of study" or the research theme of this graduate school does not point out to any specific academic discipline. Rather, I think it should be positioned as a dynamic new academic discipline that will continue to constantly revolve around the circle of core academic disciplines in the academic landscape.

5.2.3 Building Fields of Resonance Between Organizations

On the other hand, how can we reconstruct the fields of resonance which will serve as the environment for producing innovation within and between organizations?

The important places for innovation are the fields of resonance and not the organizations. However, it was Japanese society itself that has hindered paradigm disruptive innovation that is produced from the fields of resonance. As previously discussed, postwar, Japan created a social system that did not respect individual creativity.

The process of the new innovation created by the fields of resonance is inextricably linked to those fields of resonance. So, it does not apply to the paradigm of modern industry that perceives humans as a "workforce without spirit." In other words, we must redefine society as a "field of resonance that instills a desire for creation as the purpose of life."

Since fields of resonance are self-organized through the resonance of the tacit knowledge of the participants, even within the organization. It is independent of each business division of the organization.

If the creator of the fields of resonance is within an organization, the greatest challenge for the company should not be the management of knowledge by the "known school" rather than the management of fields of resonance to nurture "abduction." Although knowledge can be preserved, it is almost impossible for the "fields" to be revived once they fall apart. Therefore, it is only in the worst situation that people need to be separated and "skills shifted" within the organization; however, it is better to spin off each field of resonance.

So, what about a company split up or intrapreneurship? Unfortunately, spinning off separate companies or facilitating intrapreneurship within organizations is unlikely to increase the probability of innovations being produced.

This is because the former is simply a superficial management transformation that has made it easier to evaluate performance by each value network. Value network refers to a network of commercial systems that jointly solve problems and evolve together, based on a bond of trust for the value created between the enterprise and a customer.

The latter is merely a system that considers the individual as a property of the enterprise but shows "consideration" for one's subordinates. Therefore, the relationship between the individual and the organization does not change at all, and there is no inherent spirit to take on risks.

Moreover, as long as paradigm disruptive innovation ultimately aims to destroy existing markets and create new markets, intrapreneurship itself will be a paradox. Mental independence is a major prerequisite for a spin-off.

I would like you to recall the innovation diagram. The easiest way to produce innovation was to proceed toward "Type 0" or extend the existing paradigm. However, such paradigm sustaining innovation will always arrive at a dead-end.

When a dead-end is reached, it takes a lot of nerve, especially from the management, to discard its existing knowledge for the moment and get down to the bottom of things in the scientific horizon, which is in other words, Type 1 innovation. That is because it is not easy to go down the mountain once you have started climbing the mountain and reached the top.

When you can see the summit of the mountain, and you know for a fact that the next mountain peak is higher, it requires great courage to descend the mountain you just climbed and start scaling the other mountain again from the start. Especially, the leaders of large enterprises and nations will no doubt feel the fear of being dragged into an unknown world that they have no control over.

But this does not mean we should refrain from doing so. If managers do not have the ability to ascertain when it is time to get to the bottom of things, they should make people with such capabilities as producers and hand them the responsibility and authority to make it happen. At the same time, management should make an effort to constantly maintain fields of resonance in the workplace and acknowledge the tacit knowledge of the workplace at all times.

In addition, managers must create a circular network that ensures one to "transcend the borders of knowledge" within the fields of resonance. They must constantly maintain dynamism that overcomes the barriers of departments and circulate knowledge, especially tacit knowledge, so that Type 2 (performance disruptive innovation) and Type 3 (transdisciplinary paradigm disruptive innovation) innovations can occur.

Organizations where knowledge is stagnating and no longer communicated to the management will definitely decay. Furthermore, the creativity of organizations, where knowledge is only communicated to managers as explicit knowledge will also perish.

When such organizations encounter a situation where science has caused damage to society as in the nuclear accident, they will be petrified. They will not be able to solve the social problems that science can cause but science alone cannot solve or "trans-science" problems.

5.2.4 Bringing Scientists into the Management Team

What must be done to overcome trans-science problems and ensure scientific thinking capabilities within groups such as organizations? I think that it is important to transform organizations and include scientists in the decision-making system of organizations.

More specifically, it is important to place a chief science officer (CSO) at the core of the cross-management team who tests the source of innovation (Yamaguchi, 2007). There has been an increase in CSO positions in organizations across Europe and America.

To implement scientific thinking, some people have argued that it is enough to have a chief technology officer (CTO), the chief technical manager of technology management. However, compared to the CTO as the chief executive who has the overall responsibility for the company's technology management, the role of the CSO that I have proposed here is fundamentally different.

To explain this, I must clarify how "technology," which the CTO is expected to oversee and "science," which the CSO is accountable for are different and how they are connected.

All modern corporate activities have a core competence (a series of skill sets fundamentally supporting business capability). It is the uniqueness of this core competence that differentiates organizations from their competitors and gives them their competitive edge.

This uniqueness is made up of two elements. The first is the uniqueness of the "knowledge" itself that makes up the elements of core competence. The second is the uniqueness of the manner in which multiple "knowledge" elements are linked and integrated.

We call the intellectual activities that create the former "knowledge" as "science," and the intellectual activities by which the latter "knowledge" is integrated to give shape to value, such as products and services, as "technology."

Simply having scientists cannot be the organization's capability. Organizations must visualize the scientific thinking ability or the invisible "knowledge," give shape to the capabilities that they find, and maintain it within the organization. This is nothing but the role of the CSO.

The CTO is responsible for "knowledge embodiment" while the CSO is responsible for the "knowledge creation." These two people look at optimal allocation of employees under their supervision. Employees supervised by the CSO do not have to be present locally in the research labs.

Rather, it is preferable to always ensure scientific thinking capability in organizations by "incorporating science in organizations" and assigning the same to the Engineering Department, Safety and Security Department, CSR (Corporate Social Responsibility) Department, and so on. It is important for the CSO who is responsible for "night science" to maintain ethical tension with the CTO who is responsible for "day science." This creates a tense relationship between scientists and engineers within the organization.

Most of the Japanese corporations do not even have CTOs, let alone CSOs. Even within the manufacturing industry, CTOs are found probably only in high-tech companies. Since the general manager of JR West only uses technology, he cannot be called a CTO.

This means that the cultural and educational system that molds scientific thinking is not an integral part of organizations. The practice of getting down to the bottom of things does not lie at the foundation of the culture and thought pattern, and this problem is not an isolated issue for certain organizations but a common problem that is affecting Japanese society at large.

5.3 Toward a Society Where Everyone Pursues Science

5.3.1 Scientist Is Not an Occupation

When we reach that stage, the fields of resonance must be created not only within organizations such as universities or enterprises but also in the society itself. Specifically speaking, it is a vision where everyone is a "citizen scientist" who freely "transcends the borders of knowledge." What is a "citizen scientist?"

In order to review the relationship between citizens and science, we must look at "who is a scientist?" from a fresh perspective. First of all, let me introduce my own experiences that prove how the title of a scientist is not ingrained in the society and take this discussion forward.

This was just after I quit the NTT Basic Research Laboratory, where I had worked for over twenty years before establishing my own start-up company. In order to receive unemployment insurance, I went to a public employment security office in Tokyo.

In the midst of a severe recession, the queue, filled with mostly homeless people, extended from the office building to a small park. I took my place at the end of the queue. It was after about one hour of waiting in the queue that I finally got my turn. The first thing I learned at this counter was that this needs to be accepted as a profession first, before going ahead with the unemployment insurance formalities. An elderly official asked me:

"What is your occupation? What were you doing?"

I was flustered by the sudden question, and the words that came out of my mouth were:

"I am a scientist." The official smiled wryly on hearing my response and said:

"You know what, scientist is not an occupation. This is something a child may say when he is asked about what he wants to become in future, and not a profession. Please give me your occupation." Trying to contain my embarrassment, I answered this time: "I am a physicist."

This irritated the official even more, and he said: "Even worse."

I felt that my whole existence was being berated, so I responded vehemently: "Well, please look at the occupation classifications."

There was a thick book right in front of the officer with the title "occupation classifications." He began to turn the pages of that book, flustered by my stern expression. I blinked in amazement. This was because there was a category "physics researcher" in the first page, third row from the top.

A01 "scientist" is the top category. The first sub-category was "science researcher (011-00)," the second was "mathematician (011-01)," and the third was "physics researcher (011-03)." I think this was the first time that the official had met a "scientist" in his career of over 30 years.

I realized again how the scientist profession is disregarded by society. The "scientist" profession surfaced as a weird occupation of the 20th century industrial society. So, that is why I think citizens seem to view scientists as "people we can do without."

5.3.2   The Irresponsible Attitude of Scientists

The scientist profession is a very strange occupation (Murakami, 1994) as Yoichiro Murakami argued in "*Kagaku-sha to wa nanika? (What is a Scientist?)*."

For example, occupations such as medical doctors and engineers are valued by society as they provide specific value to society, and they are compensated by the citizens based on the value they provide.

However, because the profession of a medical doctor or a clergyman strongly implies a "vocation," the remuneration is called "honorarium" and is a "token of gratitude" that needs to be given for "shouldering the responsibility." It differs from other occupations such as a blacksmith or butcher.

However, the scientist profession is completely different from any of these occupations. The job of a scientist is "seeing things that no one has seen, knowing what no one knows, and making it exist which has never existed."

Regarding this, at a particular meeting, I had a conversation with anthropologist Juichi Yamagiwa and biologist Takahashi Yoshiko

(Kyoto Qualia Institute, 2016). Yamagiwa asked the question, "When scientists do research, what are the sources that motivate them?"

Takahashi responded: "If I am asked why I am doing such idiotic stuff, I can only say because I am drunk on my own stupidity. If what I have discovered eventually helps industry, that makes me happy. Even if it doesn't, it shouldn't matter, should it?"

In response, I added: "That is why people who do science are self-motivated, and therefore, if that motivation disappears, it would be time to quit."

Takahashi followed up saying: "No, it would be time to die."

As this argument symbolizes, scientists announce their discoveries at conferences or publish their work as papers free of charge. By doing so, the created "knowledge" is fairly transmitted far and wide around the world free of charge. Therefore, the "created knowledge" is not bought and sold and does not generate any economic value or depend on any value. This is because science is "free from value" and "value-neutral."

Scientists are never evaluated by society because they do not produce any specific value for society. Murakami calls the resulting situation as an "irresponsible attitude." Scientists advance in their careers just based on the evaluation of their colleagues. Therefore, a system where they did not have to take any responsibility toward society came into existence.

In the 19th century, "man of science" was not recognized as an occupation as scientists were indifferent to society. This changed in the 20th century, and "scientist" was established as a profession when it was positioned as a tool to produce technology, which is at the core of industry.

All of the technology that was produced in the 20th century could not be developed without relying on the science paradigm. Therefore, industrial creation based on new technology could happen only when "knowledge creation" or scientific research happened first.

5.3.3 Realization of a "Citizen Scientist Society"

I think that the only solution to find hope for science is for us citizens to go back to the original version of "person of science."

That is, to become a "person of science" or have as many people as possible who can freely transcend the borders of knowledge. This means overcoming the segmentation of academic disciplines such as humanities and science, so that anyone in different stages of life can enhance "what they think and feel" and learn "night science" to have an unobstructed view of the world.

Especially researchers in humanities/social science or lawyers and business leaders should become "persons of science" and take the initiative to evaluate the social commitment of professional scientists.

Until 19th century, there had been no term "scientists," and everyone who aspired to do science was referred to as a "person of science." Therefore, we need to go back in time to understand the original figures.

People unanimously think that science will evolve quickly and eventually split into various branches and specializations, so it will no longer be possible to have an understanding of the complete picture. However, this is incorrect. As far as science is concerned, by learning the process of "abduction" of the people who have created it, anyone can understand its essence as a life form, even if we do not understand the various techniques and terminologies.

It is my sincere hope that we can have a society of "citizen scientists," where any citizen can become a "person of science." This will also help solve the problems of trans-science relevant to both science and society.

Professional scientists and citizen scientists must understand the purpose of each other's lives and build "fields of resonance" for science and society to take on a new dimension.

Epilogue

Finally, I will summarize the recommendations of this book.

Japan has still not been able to find a new innovation model since the era of central research laboratories of large enterprises came to an end in the late 1990s. Moreover, the field of science that supports industrial competitiveness is shrinking at the same time, and it is fundamentally in a crisis.

On the other hand, the United States invented the SBIR program in 1982, and through its determined and sustained execution, they finally found a new innovation model.

This is an organic open network model by science-based start-ups, where law obliges the federal government to contribute a certain percentage of outsourced research funds to small businesses, and this was realized based on the institutional design.

In other words, it called for obscure young scientists to become innovators rather than pursue research for the sake of research and presented specific challenges toward creating new industries. Then, it encouraged applicants to become entrepreneurs of start-ups through a three-stage gate system. It is no exaggeration to say that the strong international competitiveness of science-based industries in the United States was a result of this SBIR program.

In order to prevent Japan, which is already lagging behind the rest of the world, from becoming a country where science and innovation are in danger of perishing, there is no other way but to go back to the essence of how paradigm disruptive innovation is produced and strongly support start-up companies established by the scientists.

To do so, we need to go back to the concept of the American SBIR program and create an organic open network model through science-based startups in a determined manner.

By discarding the soil, as in Japanese society that "accidentally cut off tissue while trying to cut off flab," creative young people have lost their opportunities to create and ended up becoming a working poor group. So, there is actually a lot of potential.

From this point on, I will summarize the points for reviving breakthrough innovation.

First of all, to prevent a situation where we only stick to "deduction" and consequently cannot "descend the mountain," we need to make it a practice to understand the "essence" of things by always performing "induction." Once we decide that we must get down to the bottom of things, we start descending deep into the soil. Once we go deep down into the soil, we can perform "abduction" for the first time and discover a new paradigm (Type 1: Paradigm disruptive innovation). To transform it into value and to accomplish paradigm disruptive innovation, we need to acknowledge the difference in each other's life goals and together build the "fields of resonance."

Secondly, it may not be possible to create value for the future merely by "deducing" the extension of the present. We must always visualize a "different future," "transcend the borders of knowledge" of the academic disciplines or industry, and perform "transilience" (Type 2: Performance disruptive innovation = Christensen's disruptive innovation).

Thirdly, even if we go down into the soil seeking the essence of things, we may reach a dead-end during the abduction process. But we need to try to overcome the barriers of academic disciplines and transcend the borders of knowledge (Type 3: Transdisciplinary paradigm disruptive innovation).

The future that we will have in the next ten years can only come from the current soil. Nevertheless, it is either now or never. So, we cannot afford to go back to the large enterprise central research laboratory model. We must build an exploratory-type research organization that conducts technology intelligence (exploring what kind of paradigm disruption is taking place beneath the soil), eliminate the barriers of the organization, and develop innovation sommeliers who have a good sense of judgment. Although they may not have come to our notice, there are certain Japanese start-up companies with the ability to produce paradigm disruptive innovation.

That is why there is an urgent need to build a "star discovery system" similar to SBIR of the United States in Japan. To do so, we need to create the program manager/director system, which does not exist in Japan currently, from scratch, build their career paths, and continue to develop innovation sommeliers.

We need to free ourselves from the outdated technical official system. In this age of trans-science, graduates and post-graduates absolutely lack the capability to convert scientific knowledge to social/economic value, and this is because people who have never experienced "knowledge creation" cannot conceive the future.

This program manager/director system will probably be the national chief science officer (CSO). And this will also encourage private enterprises to have a CSO.

If the social system to discover stars is put in place, alongside innovation sommeliers who explore what kind of paradigm disruption lies dormant beneath the soil and produce new innovations, we have to churn out a large number of innovators from the graduate schools.

However, it will take several years to change our mindset. It took me four years to change my mindset and convert my life goal from researcher to innovator. The first two years were a big struggle.

As part of graduate school education, teaching that there can be various kinds of mindsets, and nurturing people with a good eye for transilience will be the key to developing "fields of resonance" between knowledge creation and value creation. To that end, faculty members must have the experience of "knowledge cross-border." Graduate school reform is urgently needed.

If new graduate schools that encourage synergy between science and humanities and make innovation as their ultimate goal are set up, scientific thinking can be effectively ingrained in society. By doing this, innovation capabilities can be revived in society and in organizations, and this will also help them acquire trans-science capabilities.

References

Christensen, C. (1997) *"The Innovator's Dilemma: When New Technologies Cause Great Firms to Fail,"* Harvard Business Review Press.

Esaki, R. (2007) *"Genkai e no chōsen"* [*Challenge Your Limits*] Nikkei Publishing Inc. ISBN 978-4532166359.

Fujita, Y. (2016) Private communication, by using sbir.gov.

Fujita, Y., Kawaguchi, M., and Yamaguchi, E. (2015) *"Saiensu no fūkei - "Bunya chizu" no seisei to bunseki"* [Landscape of science - creation and analysis of the "academic landscape"] (Yamaguchi, E. "Inobēshon seisaku no kagaku" [Chapter 6, Science of innovation policy]) University of Tokyo Press, ISBN 978-4-13-046115-3.

Fukushima Project (2012) *"FUKUSHIMA Repōto - Genpatsu jiko no honshitsu"* [Fukushima Report - The true nature of nuclear accident] Nikkei BP Consulting, ISBN 978-4-86443-000-5

Hermann, A. (1976) "Werner Heisenberg in Selbstzeugnissen und Bilddokumenten".

Hiroyasu, I and Yamaguchi, E. (2015) *"Nihon no SBIR seisaku to sono kōka no beikoku tono hikaku"* [Comparison between Japan and the United States from the Standpoint of SBIR Policies and Their Effects in Japan] (Yamaguchi, E. "Inobēshon seisaku no kagaku" [Science of Innovation Policy] University of Tokyo Press, ISBN 978-4-13-046115-3).

Iijima, H. and Yamaguchi, E. (2015) "Decrease in the number of journal articles in physics in japan correlation between the number of articles and doctoral students," *J. Integrated Creative Studies*, 2015-0009, pp. 1–20.

Kawaguchi, M. (2004) *"Kaiyū-gata ijū ni kansuru ichikōsatsu - Honkon o jirei to shite"* [A Study on "Transilience" Type Migration - Case Study of Hong Kong], Soshioroji, 48(3).

Kobayashi, T. (2007) *"Toransu saiensu no jidai"* [The Era of Trans-Science], NTT Publishing Co.

Kuhn, T. S. (1962), *"The Structure of Scientific Revolutions,"* University of Chicago Press.

Kyoto Qualia Institute (2014) 10th Qualia AGORA 2014 http://www.goodkyoto.com/p005_detail.html?search=第10回クオリアAGORA 2014 / 新しい日本の針路とエネルギー

Kyoto Qualia Institute (2015) 6th Qualia AGORA 2015 http://www.goodkyoto.com/p005_detail.html?search=第６回クオリアAGORA+2015 / "無心"から+"生きる"を考える

Kyoto Qualia Institute (2015-2) 9th Qualia AGORA 2015 http://www.goodkyoto.com/p005_detail.html?search=第９回クオリアAGORA 2015 / ディスカッション

Lerner, J. (1999) "The government as venture capitalist: The long-run impact of the SBIR program," *Journal of Business*, 72(3), pp. 285–318.

Moore, G. (1965) "Cramming more components onto integrated circuits," *Electronics Magazine*, 38(8), pp. 114–117.

Murakami, Y. (1994) *"Kagaku-sha to wa nanika?"* [What is a Scientist?] Shincho Sensho, ISBN 978-4106004674.

Murakami, Y. (2010) *"Ningen ni totte kagaku to wa nanika"* [What is science for human beings?] Shincho Sensyo, ISBN 978-4-10-603662-0.

Nishihira, T. (2009) *"Zeami no keiko tetsugaku"* [Zeami's Philosophy of Practice] University of Tokyo Press, ISBN 4130101137.

Nishizawa, A. (2009) Nihon Keizai Shimbun July 23, 2009 Economics classroom.

Peirce, Charles S. (1958) *Collected Papers of Charles Sanders Peirce*, Belknap Press, Vol. 5.

Polanyi, M. (1962) "The republic of science: Its political and economic theory," *Minerva*, I(1), pp.54–73.

Richmond, A. H. (1969) "Sociology of migration in industrial and post-industrial societies" (Jackson, J. A., ed., *"Migration"*) Cambridge University Press.

Saito, M. (2014) *"Chichi ga musuko ni kataru makuro keizai-gaku"* [Macroeconomics that a father can discuss with his son] Keisōshobō.

SBIR (2016) https://www.sbir.gov

Shibatani, A. (1973) *"Han kagakuron"* [Anti-science Theory] Misuzu Shobo Co., Ltd.

Tomonaga, S. (1979) *"Butsurigaku to wa nanidarou ka?"* [What is Physics?], Iwanami sinsyo, Publishers.

Weinberg, Alvin M. (1972) "Science and trans science," *Minerva,* 10, pp. 209–222.

Wessner, C. W. (2008) *"An Assessment of the Small Business Innovation Research Program,"* National Research Council, ISBN 978-0-309-11086-0.

Yamaguchi, E. (2003) *"Handōtai debaisu sangyō"* [Semiconductor and Device Industry] (Goto, A. and Odagiri, H. *"Nihon no sangyō shisutemu 3 Saiensu-gata sangyō"* [The Japanese Industrial System - 3rd edition, Chapter 7, Science-Based Industries]) NTT Publishing Co., Ltd, ISBN 4-7571-2102-4.

Yamaguchi, E. (2006a) *"Inobēshon hakai to kyōmei"* [Innovation: Paradigm Disruptions and Fields of Resonance], NTT Publishing Co., Ltd., ISBN 4-7571-2174-1.

Yamaguchi, E. (2006b) "Rethinking innovation," Chapter 8, in *"Recovering from Success: Innovation and Technology Management in Japan,"* Oxford University Press, ISBN 978-0199297320.

Yamaguchi, E. (2007) *"JR Fukuchiyamasen jiko no honshitsu –kigyō no shakaitekisekinin o kagaku kara toraeru"* [The True Nature of JR Fukuchiyama Train Accident - Rethinking Corporate Social Responsibility from Science] NTT Publishing Co., Ltd. ISBN 978-4-7571-2196-6.

Yamaguchi, E. (2012) *"Merutodaun o fusegenakatta hontō no riyū - Fukushima daiichi genshiryokuhatsudensho jiko no kakushin"* [The real reason why meltdown could not be prevented - the core of the Fukushima Daiichi Nuclear Power Plant accident] (Fukushima Project *"Fukushima Repōto - Genpatsu Jiko No Honshitsu"* [Chapter 1, Fukushima Report - The true nature of nuclear accident]), Nikkei BP Consulting, ISBN 978-4-86443-000-5

Yamaguchi, E. (2014) *"Shinu made ni manabitai itsutsu no butsurigaku"* [Five Physics Theories to Learn Before You Die], Chikuma Sensyo, ISBN 978-4-480-01600-3.

Yamaguchi, E. (2015a) *"Inobēshon seisaku no chūkaku: SBIR seisaku to wa nanika"* [Core of the innovation policy: What is SBIR policy?] (Yamaguchi, E. *"Inobēshon seisaku no kagaku - SBIR no hyōka to mirai sangyō no sōzō"* [Chapter 1, Science of innovation policy - Evaluation of SBIR and creation of the future industry]), University of Tokyo Press, ISBN 978-4-13-046115-3

Yamaguchi, E. (2015b) *"Saiensu to toransu Saiensu"* [Science and Transscience] (Kawai, S. and Fujita, M., Chapter 20 "Advanced integrated studies in human survivability") Kyoto University Press, ISBN 978-4876988792.

Yamamoto, S. and Yamaguchi, E. (2015) *"Iyakuhin sangyō - Nihon wa naze chōraku shita ka: Inobēshon seisaku no saiteki kai"* [Pharmaceutical industry – Why there was a downfall in Japan? Optimal solution of innovation policy] (Yamaguchi, E. *"Inobēshon seisaku no kagaku - SBIR no hyōka to mirai sangyō no sōzō"* [Chapter 7, Science of innovation policy - Evaluation of SBIR and creation of the future industry]), University of Tokyo Press, ISBN 978-4-13-046115-3.

Yoshihara-Yang, M. (2015) *"Beikoku SBIR seido no genryū to rekishi"* [The origin and history of the US SBIR program] (Yamaguchi, E., *"Inobēshon seisaku no kagaku - SBIR no hyōka to mirai sangyō no sōzō"* [Chapter 2, Science of innovation policy - Evaluation of SBIR and creation of the future industry]), University of Tokyo Press, ISBN 978-4-13-046115-3.

Index